Mastering Sheet Metal Design with Autodesk Inventor™

by Travis Jones and Neil Munro

DELMAR
THOMSON LEARNING™

Mastering Sheet Metal Design with Autodesk Inventor™

Travis Jones and Neil Munro

Executive Director
Alar Elken

Executive Editor
Sandy Clark

Senior Acquisitions Editor
James Devoe

Senior Development Editor
John Fisher

Executive Marketing Manager
Maura Theriault

Channel Manager
Fair Huntoon

Marketing Coordinator
Sarena Douglass

Executive Production Manager
Mary Ellen Black

Production Manager
Andrew Crouth

Production Editor
Stacy Masucci

Senior Art/Design Coordinator
Mary Beth Vought

Technology Project Manager
David Porush

Technology Project Specialist
Kevin Smith

Editorial Assistant
Mary Ellen Martino

Printed in the United States of America
1 2 3 4 5 XX 06 05 04 03 02

For more information contact
Delmar Learning
Executive Woods
5 Maxwell Drive, PO Box 8007,
Clifton Park, NY 12065-8007
Or find us on the World Wide Web at
http://www.delmar.com

Library of Congress Catalog data

ISBN: 1-4018-2677-6

NOTICE TO THE READER

Publisher does not warrant or guarantee any of the products described herein or perform any independent analysis in connection with any of the product information contained herein. Publisher does not assume, and expressly disclaims, any obligation to obtain and include information other than that provided to it by the manufacturer.

The reader is expressly warned to consider and adopt all safety precautions that might be indicated by the activities herein and to avoid all potential hazards. By following the instructions contained herein, the reader willingly assumes all risks in connection with such instructions.

The publisher makes no representation or warranties of any kind, including but not limited to, the warranties of fitness for particular purpose or merchantability, nor are any such representations implied with respect to the material set forth herein, and the publisher takes no responsibility with respect to such material. The publisher shall not be liable for any special, consequential, or exemplary damages resulting, in whole or part, from the readers' use of, or reliance upon, this material.

Table of Contents

Welcome ix

Included Files on the CD. ix

Project File for Exercises x

Support Information xi

Your Autodesk Authorized Manufacturing Reseller. . . .xi

United States and Canada xi

Latin America and Caribbean. xi

Direct Support Options. xi

User Groups xi

Free Support xi

Prerequisites xii

Acknowledgments. xii

Dedications xii

Mastering Sheet Metal Design with Autodesk Inventor 1

Course Objectives 1

Introduction to Sheet Metal Design 2

Sheet Metal Fabrication 2

Sheet Metal Parts 3

Sheet Metal Design Methods. 3

Creating Sheet Metal Parts in Autodesk Inventor . . 3

Sheet Metal Tools 4

Sheet Metal Styles 4

Sheet Tab. 6

Bend Tab 7
Corner Tab 8
Face 9
Contour Flange 11
Flange 13
Hem 14
Fold 16
Bend 17
Corner Seam 18
Cut 20
Corner Round 21
Corner Chamfer 22
Punch Tool 23
Flat Pattern 24
Common Tools 25
Detailing Sheet Metal Designs 25
Exercise 1: Sheet Metal in Assemblies **27**
Cross Part Associativity 27
In-Place Part Creation 27
Add Sheet Metal Faces 30
Adding Sheet Metal Features 35
Add Corner Seams 38
Flat Pattern Creation 40
Adaptive Sheet Metal Update 42
Detail the Sheet Metal Part 44
Add a Folded Model View 45
Face and Flange Techniques **47**
Bend Tables **48**
Bend Table Editing 50
Bend and Corner Reliefs 51
Exercise 2: Creating Faces and Flanges **53**
Creating a Bend Table 53
Sheet Metal Style Setup 55
Update the Sheet Metal Style 55

Creating Faces and Bends 57
Creating Bends using the KFactor 59
Add an Additional Flange 61
Exercise 3: Rolled Sheet Metal Parts 64
Working with Flat Patterns 65
Exercise 4: Working with Imported Geometry . . . 66
Open an AutoCAD Drawing 66
Flat Shape 67
Fold Features 68
Add Fold Features 69
Add Additional Folds 72
Exercise 5: Importing a Folded Model 74
iFeatures and Punches 77
PEM Nut Relief 77
Method A 77
Method B 77
Method C 78
Method D 78
Exercise 6: iFeatures and Punches 79
PEM Nut Hex Clearance iFeature 79
Add Additional Sketch Features 80
Extract the iFeature 82
Bridge Lance Punch 82
Extract the iFeature 85
Placing Punch iFeatures 85
Place the Sheet Metal Punch 86
Sheet Metal Parts from Solids 88
Exercise 7: Shelled Solids 89
Rip the Corners 90
Add Bends 92
Add Corner Seams 93
Exercise 8: Lofted Shapes 95
Derive a Component 96

Update the Sheet Metal Style 98
Add Sheet Metal Faces. 98
Create a Second Sheet Metal Part 101
Derive a Parameter 102
Derive a Second Component 102
Add Sheet Metal Faces. 104
Add Sheet Metal Flanges 106
Applying Your Skills 110
Course Summary. 111

Welcome

Welcome to the Autodesk® Manufacturing Courseware for Mastering Sheet Metal Design with Autodesk Inventor. This courseware includes a Student Guide and a set of data for performing the exercises described within the Student Guide. To get started, read the following two sections entitled *Included Files on the CD* and *Project File for Exercises.*

Included Files on the CD

Each book ships with a CD containing files that will be needed to complete the exercises. It is recommended that you run setup from the CD and install the files in the default directory. The default installation directory is \Program Files \Autodesk \Mastering Sheet Metal Design.

The exercises included in the Design Support System (DSS) Volume, which will be installed, are the same as those included in the book. If you prefer to work electronically you can use the DSS Volume, by launching it from the shortcut that will be installed on your desktop, to complete the exercise steps while working with Autodesk Inventor at the same time. Open the DSS Volume and then resize the Autodesk Inventor application and position it next to the DSS Volume as shown.

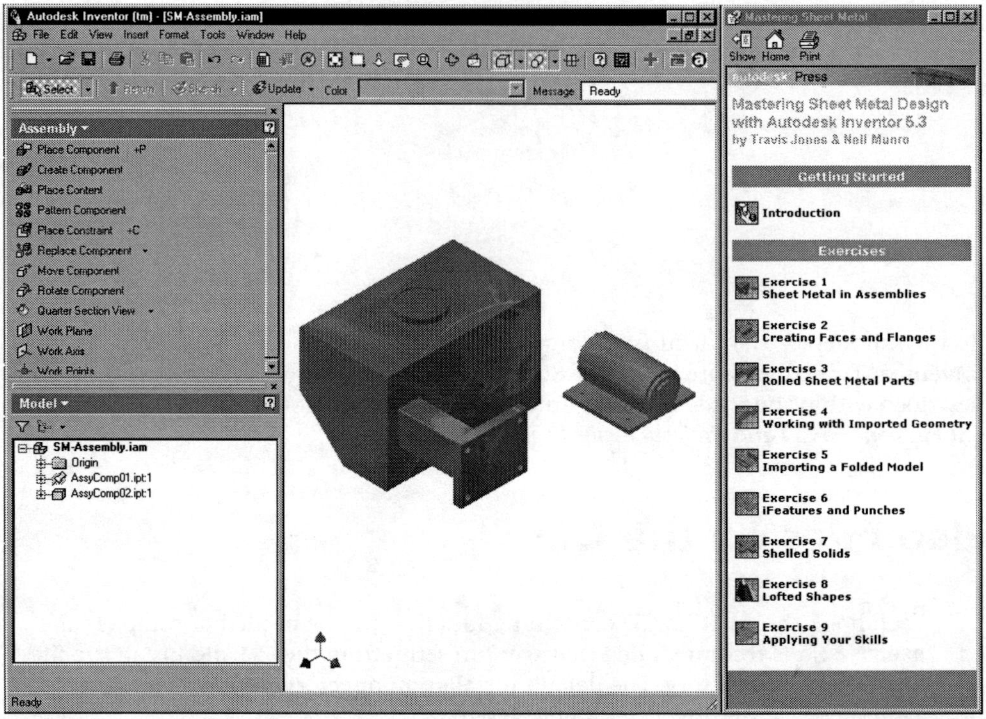

Project File for Exercises

After running the installation routine, you need to set the *Mastering Sheet Metal.ipj* file as the active project. To do this:

1. Start Autodesk Inventor.
2. Click *Projects* in the What To Do list on the left of the Startup dialog box.
3. Click the Browse button at the bottom of the dialog box.
4. Navigate to the folder where you installed the training materials then double-click the *Mastering Sheet Metal.ipj* file.
5. In the Projects window, double-click the Mastering Sheet Metal project to make it the active project.
6. Click Cancel at the bottom of the Startup dialog box.

You are now ready to work with the exercises from either the printed manual or the DSS Volume.

Support Information

Autodesk supports your product through the following channels:

Your Autodesk Authorized Manufacturing Reseller

Autodesk Authorized Manufacturing Resellers provide basic installation support. Technical user support is not included but may be purchased for a fee. You can locate Autodesk resellers as follows:

United States and Canada

Phone: 800-964-6432

Latin America and Caribbean

Fax: +1-415-507-6110

Or you may call your local Autodesk office.

Direct Support Options

Detailed information about web, e-mail, and telephone support options in your area can be found in the Support Resources locator at **www.autodesk.com/support-resources**.

User Groups

Local Autodesk user groups meet in many places around the world. Autodesk also supports the Autodesk User Group International (AUGI®), a nonprofit professional association of Autodesk product users. For more information, visit **www.augi.com** (membership required).

Free Support

At **www.autodesk.com/support** you'll find:

- Knowledge base: Find the answers to your technical questions in this knowledge base of Autodesk technical support information
- Downloads: Get the latest patches, updates, drivers, and utilities for your product
- Discussion groups: Ask questions or share technical information and tips with other users and with Autodesk staff in this moderated forum
- BigFix™: Get free, real-time, proactive technical support

Prerequisites

It is recommended that you have a working knowledge of Microsoft Windows 98, Windows NT 4.0/Windows 2000, or Windows XP, and a working knowledge of parametric solid modeling concepts. You should also have attended an introduction to Autodesk Inventor course or have a good understanding of how to create parts, assemblies, and drawings using Autodesk Inventor.

Acknowledgments

This course would not have been possible without a contribution from many people.

The authors would like to send a special thanks to the following individuals:

- Kent Keller, Strobe Data Inc. – Designer
- Jason Ruge, Autodesk, Inc. – Manufacturing Product Support
- Rockie Lyons Beaman, Autodesk, Inc. – Manufacturing Project Manager
- John Fisher, Delmar Thomson Learning – Developmental Editor
- James DeVoe, Delmar Thomson Learning – Acquisitions Editor
- Stacy Masucci, Delmar Thomson Learning – Production Editor

Dedications

To my kids, Jake and Dakota, who supported this opportunity and willingly sacrificed their time. To my wife, Kelli, for her love and steady strength, which keeps me going. Also, to my parents who encouraged my love for computers and CAD.

—Travis Jones

Mastering Sheet Metal Design with Autodesk Inventor

This course is designed to provide a thorough understanding of how to create and use sheet metal parts in Autodesk Inventor. You will learn how sheet metal parts are designed and used in industry, review various construction techniques, how to master and automate sheet metal design tools, and how to document sheet metal designs.

Course Objectives

Demonstrate how sheet metal parts are designed and used, including

- Enclosures
- Brackets
- Structures or Frames
- Formed Components

Describe the various construction techniques, including

- Face – Flange construction
- Shelled Solids with Ripped Corners
- Flat pattern to folded part
- Imported geometry
- In-place assembly design using adaptive design techniques

Optimize various sheet metal tools, including

- Sheet Metal Styles
- Bends, Flanges, Hems, Corner Seams, Cuts and Folds
- Flat Patterns

Automate sheet metal tools, including

- Creating Sheet Metal iFeatures
- Punch tool

Demonstrate how to detail sheet metal designs.

Introduction to Sheet Metal Design

A metal blank folded into a finished shape is a common component in a mechanical assembly. Examples include enclosures, guards, simple to complex bracketry, and structural members. Although the term sheet metal is often associated with these components, heavy plates (1"+) can also be formed using similar methods.

Common sheet metal components include electronic and consumer product chasses and enclosures, lighting fixtures, support brackets, frame components, and drive guards.

Sheet metal fabrication can include a number of processes.

- Drawing
- Stamping
- Punching and cutting
- Braking
- Rolling
- Other more complex operations

The tools in the Autodesk Inventor sheet metal environment enable you to create press brake formed models. You can also create rolled shapes including cones and cylinders. In addition, sheet metal parts can include formed features such as nail holes, lances, and louvers. Using lofts and surfacing tools, you can create parts that are fabricated through deformation processes such as drawing or stamping, but Autodesk Inventor cannot unfold these parts to determine the shape of the flat blank.

Sheet Metal Fabrication

Brake formed parts must be described in a minimum of two states, the flat blank shape prior to bending, and the finished part in its folded form. Manufacturing processes are often performed on the flat sheet before folding the part into a finished shape. Holes and other openings are punched into the flat stock, and deformations such as dimples are stamped into the blank with forming tools and dies. Preparing the flat stock can be done manually, or more commonly with CNC punching machines. High volume sheet metal parts are often created on progressive die lines where a continuous feed of flat stock (from a coil) is passed through a series of forming dies that remove material or form the shape of the component. For lower volume components, manual or CNC press brakes are the primary tools used to bend the finished blank into its bend up form.

The following exercises and tool descriptions outline the various workflows for creating sheet metal parts in Autodesk Inventor.

Sheet Metal Parts

In this section, you will learn how to create sheet metal parts in Autodesk Inventor. Assemblies often require components that are manufactured by bending flat metal stock to form brackets or enclosures. Cutouts, holes, and notches are cut or punched from the flat sheet and 3D deformations such as dimples or louvers are often formed into the flat sheet. The punched sheet is then bent at specific locations using a press brake, or other forming tools, to create a finished part.

Sheet Metal Design Methods

Sheet metal parts are most often created in the folded state. The model is then unfolded into a flat sheet using the Flat Pattern tool. To create the folded model, the first sketch is extruded the thickness of the sheet to create a face. Additional faces or flanges are added to open edges of the part and bends are automatically added between the features. Cuts and other special sheet metal features are added to the part as required.

There may be some instances where only the flat pattern dimensions of a sheet metal part are known. Use the Fold tool to add bends to a face that represents a flat pattern.

Sheet metal parts can also be created from standard parts that have been shelled. All walls of the shelled part must be the same thickness, and match the thickness of the flat sheet. The solid corners of the shelled part are ripped open to enable the model to unfold, and appropriate bends are added along edges between faces.

Autodesk Inventor's ability to create models with disjointed solids (unconnected features) is very helpful when building sheet metal parts in an assembly. Separated faces of a sheet metal part can be built quickly by referencing faces on other parts in the assembly. Additional sheet metal features can then join the distinct faces.

Creating Sheet Metal Parts in Autodesk Inventor

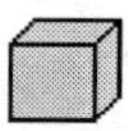

Sheet Metal.ipt

The first step in creating a sheet metal part is to either select the New icon from the What To Do section, or select New from the pull down menu on the Standard toolbar, and then select the Sheet Metal.ipt icon from the Default template tab. If the *Standard.ipt* file is used, the sheet metal tools can be loaded by selecting Sheet Metal from the Applications pull-down menu as shown in Figure 1-1.

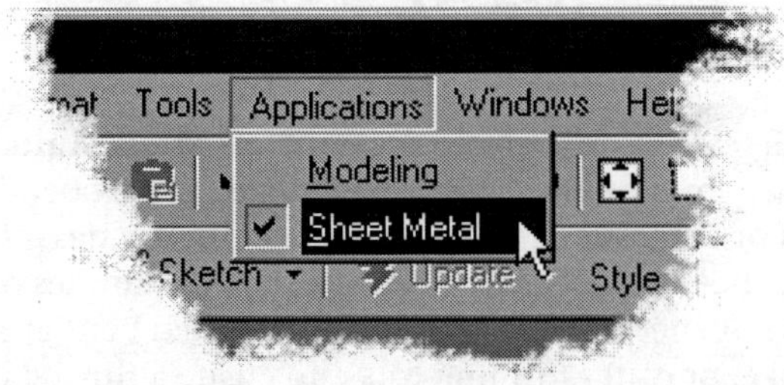

Figure 1-1

Sheet Metal Tools

After issuing the new sheet metal tool, Autodesk Inventor's screen will change to reflect the sheet metal environment. As with standard parts, a sketch is created and is active. The sketch tools are common to both sheet metal and standard parts. The first sketch of a sheet metal part must either be a closed profile that will be extruded a distance equal to the sheet thickness to create a sheet metal face, or an open profile that will be thickened and extruded as a contour flange. Additional tools are used to add sheet metal features to the base feature. The Sheet Metal Panel Bar, which contains all of the sheet metal tools, is shown in Figure 1-2.

The following paragraphs are a description of the tools used to create sheet metal features. You can also add standard part features, such as Extrude and Revolve, to a sheet metal part, but these features may not unfold when a flat pattern is generated from the folded model.

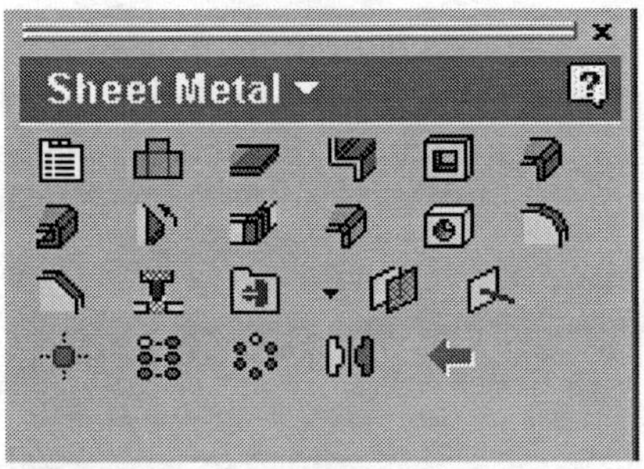

Figure 1-2

Sheet Metal Styles

The sheet metal specific parameters of a part are stored in a sheet metal style. These include the thickness and material of the sheet metal stock, a bend allowance factor to account for metal stretching during the creation of bends, and various parameters dealing with sheet metal bends and corners. You can create additional sheet metal styles to account for various materials and manufacturing processes or material types. If you create sheet metal styles in a template file, they will be available in all sheet metal parts based on that template.

The thickness of the sheet metal stock is the key parameter in a sheet metal part. Sheet metal tools such as face, flange, and cut automatically use the Thickness parameter to ensure that all features are the same wall thickness, a requirement for unfolding a model. In the default sheet metal style, all other sheet metal parameters such as bend radius are based on the Thickness value.

To edit or create a sheet metal style, follow these steps:

1 Select the Styles tool from the Sheet Metal Panel Bar. The Sheet Metal Styles dialog box will appear as shown in Figure 1-3.
2 To create a new sheet metal style, click the New button in the Sheet Metal Styles dialog box. A copy of the style currently displayed in the dialog box is created.
3 Edit the values on the Sheet, Bend, and Corner tabs to define the feature properties of parts created with this sheet metal style.
4 Rename the style and click the Save button.
5 When multiple sheet metal styles exist in a part, set the active style by selecting it from the Active Style list.

Changing to a different sheet metal style, or making changes to the active style, updates the sheet metal part to match the new settings.

The following tables contain a description of the settings in a sheet metal style. The parameters with numeric values, such as Thickness and Bend Radius, are saved as model parameters. The parameter values are updated when a sheet metal style is modified, or a new sheet metal style is activated. Other model parameters, and user-defined parameters can reference these parameters in equations.

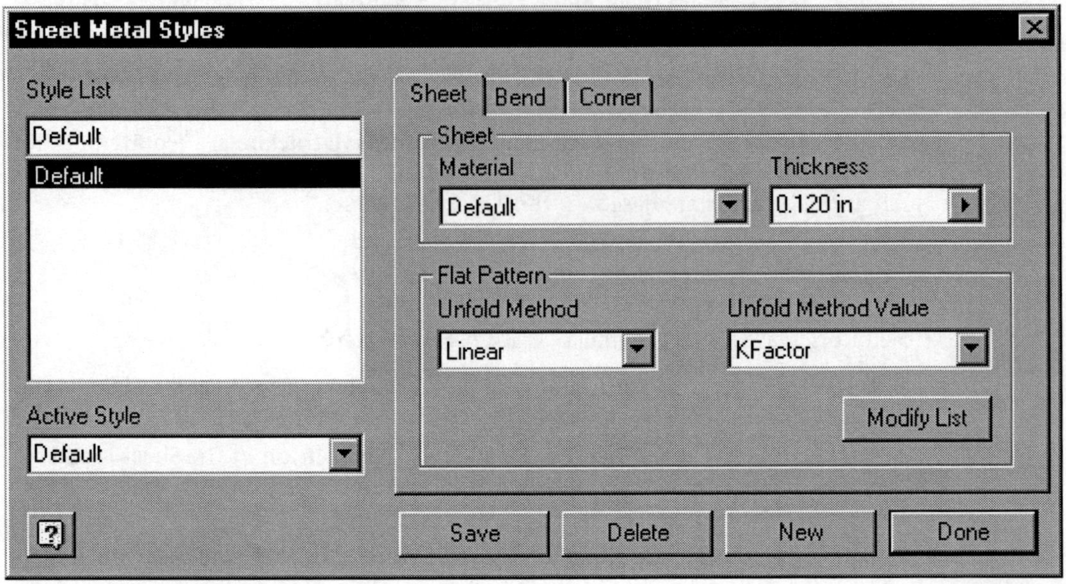

Figure 1-3

Sheet Tab

See Figure 1-3.

Material Select from a list of defined materials in the document. The material color setting is applied to the part.

Thickness Thickness of the flat stock that will be used to create the sheet metal part.

Unfold Method Select the method used to calculate bend allowance. Bend allowance accounts for material stretching during bending. Options are a constant KFactor, or a bend table for more complex requirements.

Unfold Method Value Combined with Unfold Method. The displayed value is the default value used for calculating bend allowances. A KFactor is a value between 0 and 1 that indicates the relative distance from the inside of the bend to the neutral axis of the bend. A KFactor of 0.5 specifies the neutral axis lies at the center of the material thickness.

Modify List Maintains a list of KFactors and named Bend Tables included in the sheet metal style. Each bend can have the default bend allowance overridden by a value in this list.

Bend Tab

See Figure 1-5.

Radius The inside bend radius between adjacent, connected faces.

Minimum Remnant If a bend relief cuts within this distance the end of an edge, the small tab of remaining material is also removed.

Transition Controls the intersection of edges across a bend in the flattened sheet. For bends without bend relief, the unfolded shape is a complex surface. Transition settings simplify the results creating straight lines or arcs, which can be cut in the flat sheet before bending. Samples of the four transition types are shown in Figure 1-4.

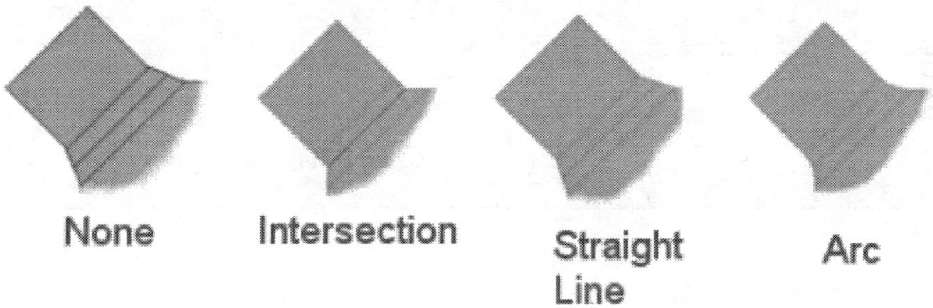

Figure 1-4

Relief Shape If a bend does not extend the full width of an edge, a small notch is often cut next to the end of the edge to keep the metal from tearing at the edge of the bend.

Relief Width Width of the bend relief.

Relief Depth Setback amount from the edge. Round relief shapes require this be at least one half the Relief Width.

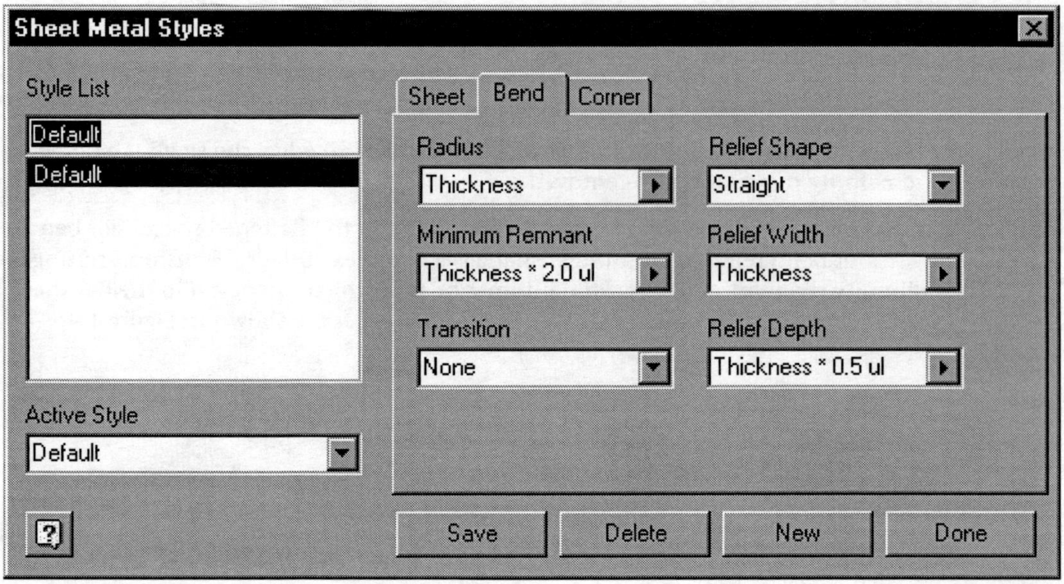

Figure 1-5

Corner Tab

See Figure 1-6.

A corner occurs where three faces meet. To enable efficient bending, a small amount of material is often removed from the flat sheet at the location where the three faces intersect. You can choose from four corner relief options and the shape of the corner relief can be viewed in the flat pattern of the model. Sample corner relief shapes are shown in Figure 1-7.

Relief Shape — Specifies the shape of the corner relief.

Relief Size — Sets the size of the corner relief when the shape is set to round or square.

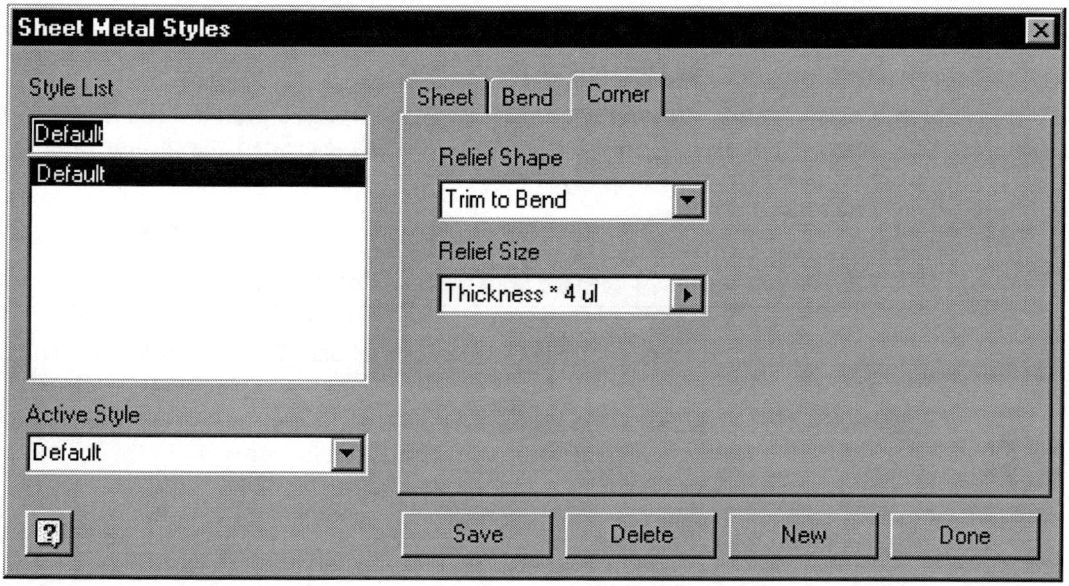

Figure 1-6

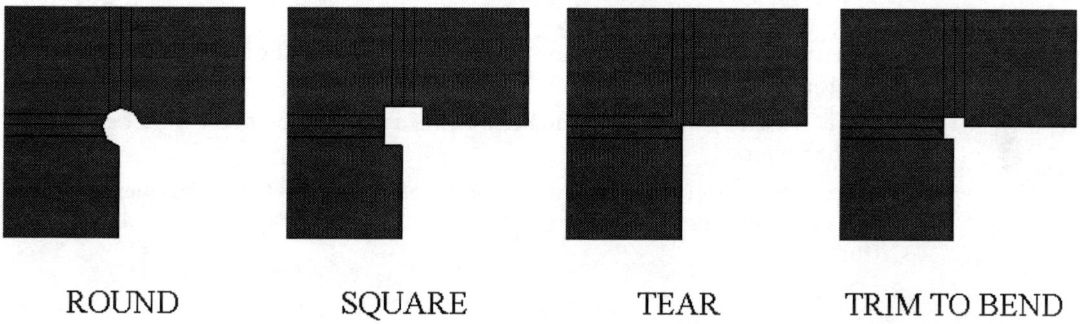

Figure 1-7

Face

The Face tool extrudes a closed profile a distance equal to the sheet metal thickness. If the face is the first feature in a sheet metal part, you can only flip the direction of the extrusion. When a face is created later in the design, an adjacent face can be connected to the new face with a bend. If the sketch shares an edge with an existing feature, the bend is automatically added. You can optionally select a parallel edge on a disjointed face. This will extend or trim the attached face to meet the new face, with a bend created between the two faces.

To create a sheet metal face, follow these steps:

1. Create a sketch with a single closed profile or a closed profile containing islands. The sketch is most often on a work plane created at a specific orientation to other part features or created by selecting a face on another part in an assembly.

NOTE You can create a single face from multiple closed profiles.

---

2. Select the Face tool from the Sheet Metal Panel Bar. The Face dialog box will appear as shown in Figure 1-8.
3. If a single closed profile is available, it is selected. If multiple closed profiles are available, you must click inside one to define the face area.
4. If required, flip the side to extrude the profile.
5. If the face is not the first feature and the sketch does not share an edge with an existing feature, click the Edges button and select a parallel edge. The two faces are extended or trimmed as required, meeting at a bend. The bend is listed as a child of the new face in the browser. If the face attached to the selected edge is parallel but not coplanar with the new face, a set of double bend options are presented. An additional face is added to connect the two parallel faces. The orientation and shape of this face is determined by the selected double bend option.
6. Clear the checkmark from the Bend Relief option if you don't want bend reliefs to be included.
7. The Unfold Options and Bend Options tab contain settings for overriding the default values in the sheet metal style. Overrides are applied to individual features.
8. Click OK.

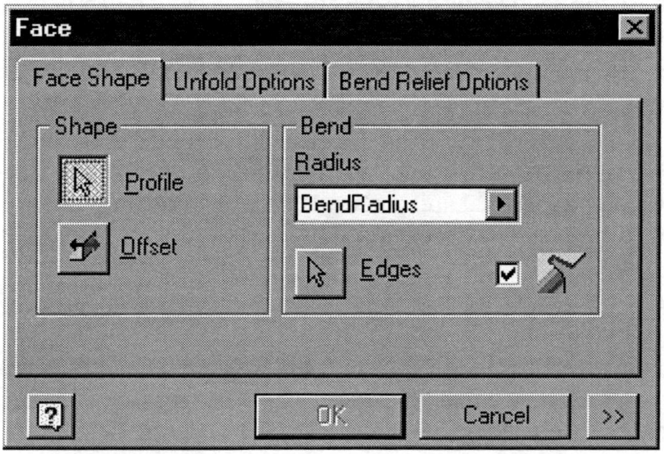

Figure 1-8

Contour Flange

A contour flange is created from an open sketch profile. The sketch is extruded perpendicular to the sketch plane and the open profile is thickened to match the sheet metal thickness. The profile does not require a sketched radius between line segments; bends are added at sharp intersections. Arc or spline segments are offset by the sheet metal thickness. A contour flange can be the first feature in a sheet metal part, or it can be added to existing features.

To create a contour flange, follow these steps:

1 Create a sketch with a single open. The sketch can contain line, arc, and spline segments. The sketch can be constrained to projected reference geometry to define a common edge between the contour flange and an existing face.
2 Select the Contour Flange tool from the Sheet Metal Panel Bar. The Contour Flange dialog box will appear as shown in Figure 1-10.
3 Select the open sketch profile to define the shape of the contour flange.
4 If required, flip the side to offset the profile.
5 If the contour flange is the first feature in the part, enter an extrusion distance and direction. If the contour flange is not the first feature, an edge perpendicular to the sketch plane can be selected to define the extrusion extents. If the sketch is attached to an existing edge, select that edge. If the sketch is not attached to any projected geometry, the contour flange is extended or trimmed to meet the selected edge. The selected edge is typically the closest edge to the end of the sketch.

6 Uncheck the Bend Relief checkbox to exclude bend reliefs between the contour flange and the selected edge.

7 Apply any unfold and bend overrides on the Unfold Options and Bend Relief Options tabs.

8 If an edge is selected to define the extrusion extents, you can select from three options to further refine the length of the contour flange. The extent types below can be specified and are available when creating other types of sheet metal features such as flanges and hems (samples of these options are shown in Figure 1-9):

Edge — The contour flange extends the full length of the selected edge.

Width — A point defines one extend of the contour flange. The flange can be offset from this point and extends a fixed distance. The starting point is defined by selecting an endpoint of the selected edge, a work point on a line defined by the edge, or a work plane perpendicular to the selected edge.

Offset — Similar to the Width option, two points are selected that define both extents of the flange. An offset distance from each point can be specified.

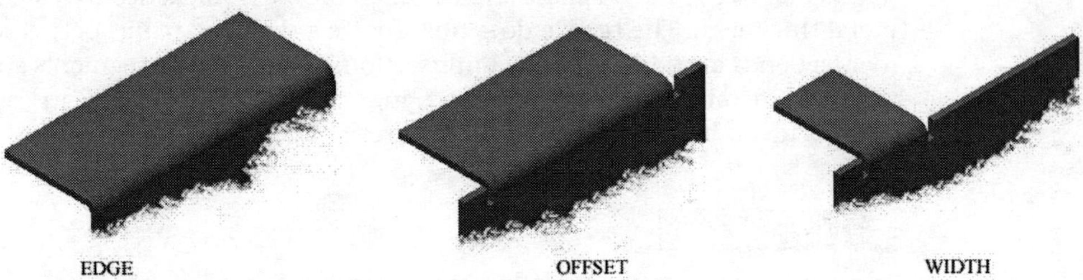

Figure 1-9

9 Click OK.

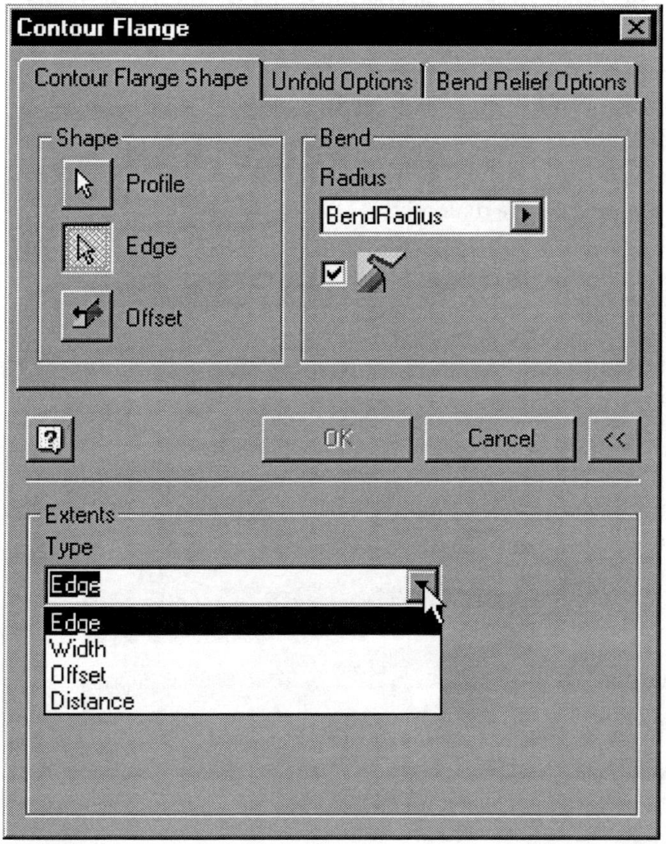

Figure 1-10

Flange

The flange tool creates a sheet metal face and a bend to an existing face. The selected edge defines the bend location between the two faces.

To create a flange, follow these steps:

1. A minimum of one sheet metal face must exist.
2. Select the Flange tool from the Sheet Metal Panel Bar. The Flange dialog box will appear as shown in Figure 1-11.
3. Select an edge on an existing face.

4 Enter the distance and angle for the flange in the Flange dialog box. The flange preview updates to match the current values.
5 If required, flip the direction of offset side for the flange.
6 Expand the Flange dialog box and select the appropriate extents type. See Contour Flange for additional information on extents.
7 Apply the Flange to continue creating Flanges, or click OK to complete the Flange and exit the dialog box.

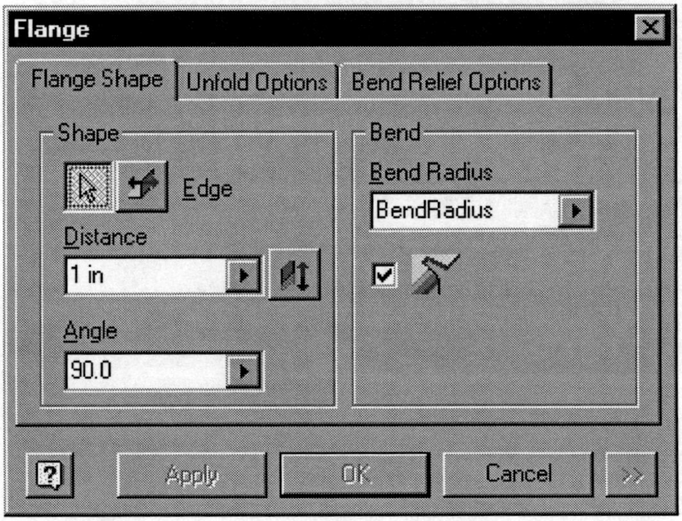

Figure 1-11

Hem

Hems are used to eliminate sharp edges or strengthen an open edge of a face. Material is folded back over the face with a small gap between the face and the hem. A hem does not change the length of the sheet metal part; the face is trimmed so the hem is tangent to the original length of the face.

To create a hem, follow these steps:

1 A minimum of one sheet metal face must exist.
2 Select the Hem tool from the Sheet Metal Panel Bar. The Hem dialog box will appear as shown in Figure 1-13.
3 Select an open edge on a sheet metal face.
4 Select the hem type (samples are shown in Figure 1-12):

- Single—A 180° flange.
- Double—Single hem is folded 180°, resulting in a double thickness hem.
- Teardrop—A single hem in a teardrop shape.
- Rolled—A cylindrical hem.

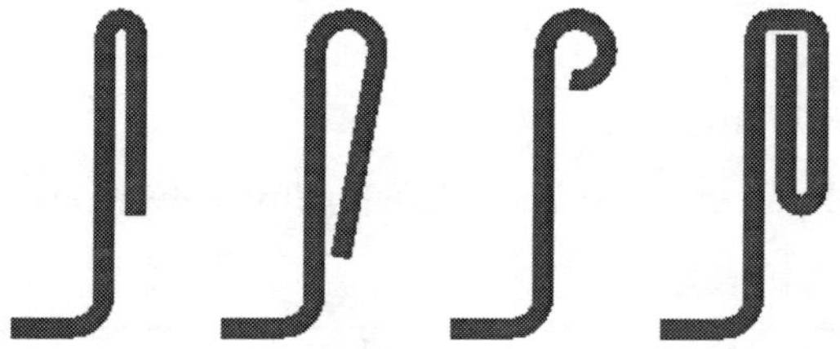

Figure 1-12

5 Enter values for the hem. Teardrop and rolled hems require radius and angle values, single and double hems require a gap and length values. The hem preview changes to match the current values.

6 Expand the dialog box and select Edge, Width, or Offset for the hem extents. See Contour Flange for additional information on defining extents.

7 Apply the hem to continue creating hems, or click OK to complete the hem and exit.

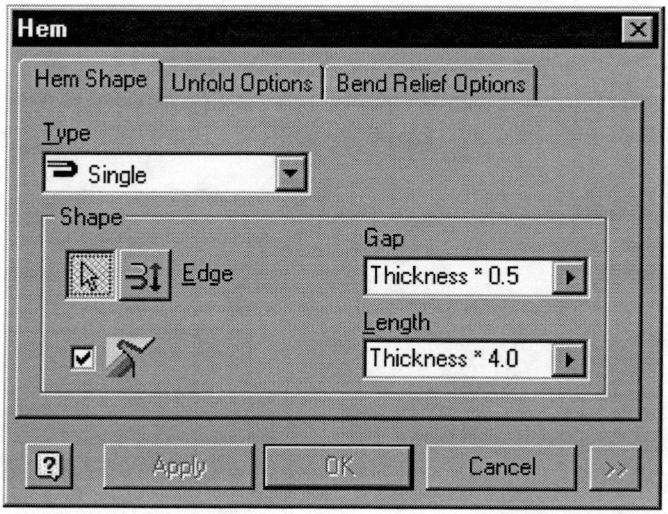

Figure 1-13

Fold

An alternate method for creating sheet metal features is to start with a known flat pattern shape and add folds at sketched lines on a face. The Fold tool can create sheet metal shapes that are difficult to create using Face or Flange tools.

To create a fold, follow these steps:

1 Create a sketch on an existing face. Sketch a line between two open edges on the face.

NOTE The line endpoints must be coincident to the face edge.

---

2 Select the Fold tool from the Sheet Metal Panel Bar. The Fold dialog box will appear as shown in Figure 1-14.

3 Select the sketch line. The fold direction and angle are previewed in the graphics window. The fold arrows extend from the face that will remain fixed. The face on the other side of the bend line will fold around the bend line.

4 If required, flip the fold direction and side.

5 Enter the angle of the fold.

6 Select the positioning of the fold with respect to the sketched line. The line can define the start, centerline, or end of the bend. The fold preview updates to match the current settings.

7 Make any changes to unfold or bend relief options.

8 Apply the fold to continue creating folds (an unconsumed sketch with valid geometry must be available), or click OK to complete the fold and exit.

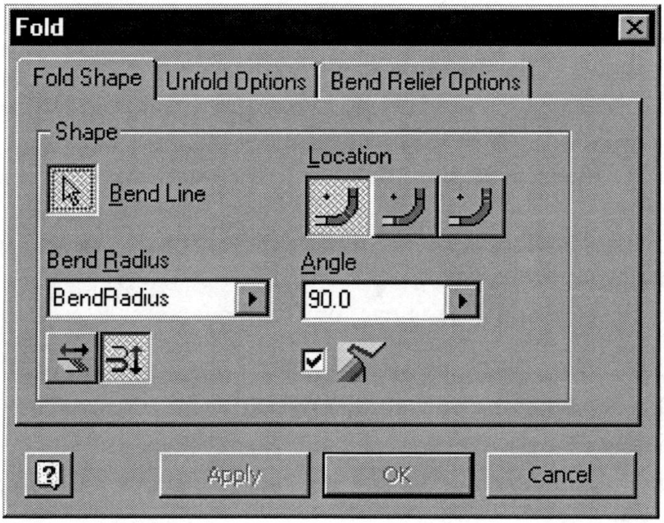

Figure 1-14

Bend

Bends are created as child objects of other features when the feature results in two faces being connected. Bends can also be created as independent objects between disjointed faces. A preview of the bend is shown and the faces that are being joined are either trimmed or extended based to connect the faces.

When parallel faces are selected for the bend feature, a jog or z-bend will be created depending on the edge location. Jogs are often used to allow overlapping material. Double bends can also be created when the two faces are parallel and the selected edges face the same direction. You can create the bend to be two 90° bends with a tangent face between the bends or specify a full radius bend between the two faces as shown in Figure 1-15.

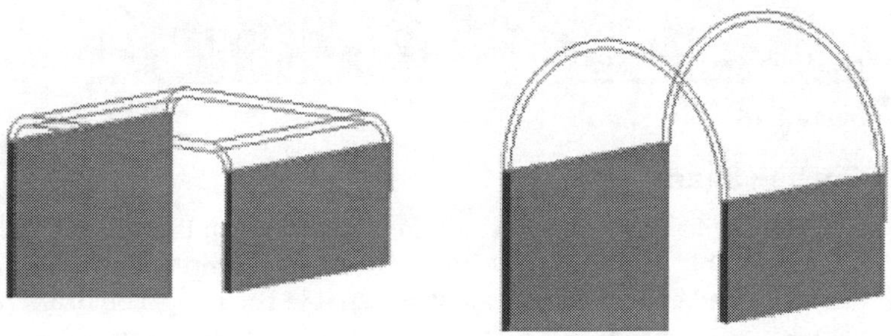

Figure 1-15

To create a bend, follow these steps:

1. The sheet metal part must have two disjointed or intersecting faces. An example of intersecting faces is a shelled box that is being changed into a sheet metal part. The intersection of the box base and a wall is an edge that can be changed to a bend.
2. Select the Bend tool from the Sheet Metal Panel Bar. The Bend dialog box will appear as shown in Figure 1-16.
3. Select the common edge of two intersecting faces, or select two parallel edges on disjointed, non-coplanar faces. If the two faces are parallel, a set of Double Bend options is presented. Depending on the position of the two faces, the Double Bend options will be either 45° and Fixed Edges, or Full Radius and 90°.
4. Make any changes to unfold or bend relief options.
5. Apply the bend to continue creating bends, or click OK to complete the Bend and exit.

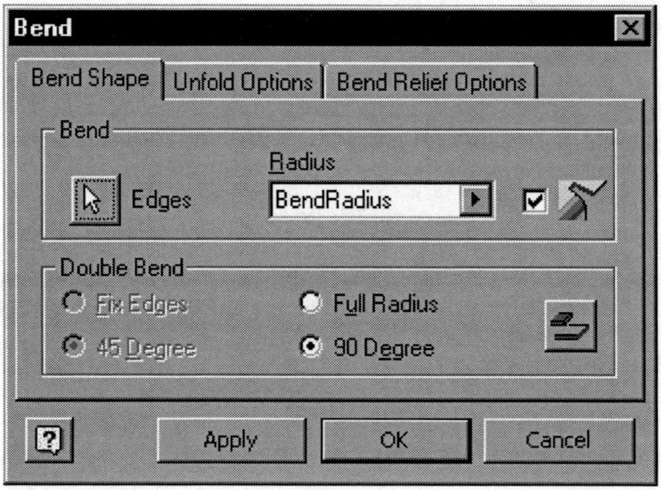

Figure 1-16

Corner Seam

When three faces meet in a sheet metal part, a gap is required between two of the faces to enable unfolding. Using a box as an example, the walls are connected to the floor, and gaps between the walls enable the box to be unfolded. The gap between adjacent faces is a corner seam. Corner seams can also be used to create mitered gaps between coplanar faces. Finally, a corner seam can be created by "ripping" open a corner at an intersection between faces.

To create a corner seam, follow these steps:

1 Select the Corner Seam tool from the Sheet Metal Panel Bar. The Corner Seam dialog box will appear as shown in Figure 1-17.

2 Select face edges that meet one of the following:

- Two open edges on faces that share a common connected edge. (Two walls that share a connection to a floor.)
- Two nonparallel edges on coplanar faces (creates a mitered joint with a gap between the extended faces.)
- A single edge at the intersection of two faces. (The wall intersections of a shelled box).

3 Select the Miter option in the Corner Seam dialog box.

4 Enter a value for the corner seam gap.

5 Make any changes to the corner options.

6 Apply the corner seam to continue creating corner seams, or click OK to complete the corner seam and exit.

NOTE The actual sheet metal thickness must equal the thickness in the sheet metal style for the corner seam operation to work.

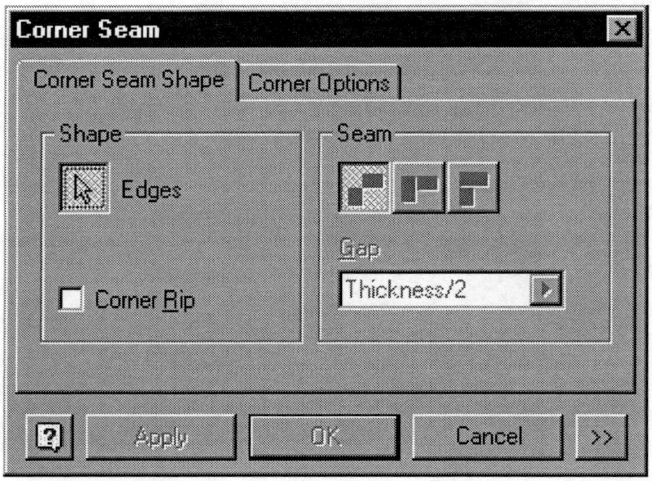

Figure 1-17

Cut

The Cut tool is a sheet metal specific implementation of the standard extrude tool. The Cut tool always performs an extrude cut; join and intersection are not available. The default extents are a blind cut a distance equal to the sheet metal Thickness parameter. This ensures that the cut only extends through the face containing the sketch, and not through other faces that may be folded under the sketch face.

Most cuts are made on the flat sheet stock before the sheet is bent to form the folded part. The cuts often cross bend lines. Because the part is modeled in the folded state, representing cuts that cross bend boundaries requires the bend be unfolded to represent the flat sheet. When the bend is refolded, the cut deforms around the bend to ensure that the extrusion remains perpendicular to the sheet metal faces on both sides of the bend. When sketching a cut profile that will cross a bend, the Project Flat Pattern tool projects the unfolded flat pattern geometry onto the sketch face.

To create a cut feature, follow these steps:

1. Create a sketch on a sheet metal face that includes one or more closed profiles representing the area(s) to cut. If required, use the Project Flat Pattern tool to project the unfolded geometry of connected faces onto the sketch.
2. Select the Cut tool from the Sheet Metal Panel Bar. The Cut dialog box will appear as shown in Figure 1-18.
3. Select the profiles to cut.
4. If a profile crosses a bend, check the Cut Across Bend option in the Cut dialog box.
5. Apply the cut to continue creating cuts (an unconsumed sketch must be available), or click OK to complete the cut and exit.

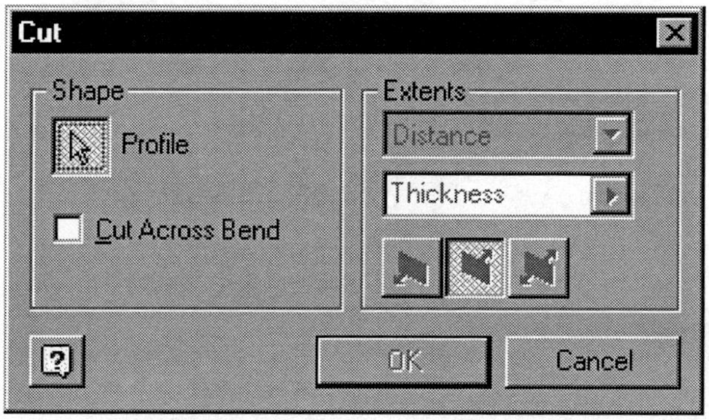

Figure 1-18

Corner Round

The Corner Round tool is a sheet metal specific fillet tool. All edges other than the ones at open corners of faces (all of these edges have a length = Thickness) are filtered out. This enables you to easily select these small edges without zooming in on the part.

To create a corner round, follow these steps:

1 Select the Corner Round tool from the Sheet Metal Panel Bar. The Corner Round dialog box will appear as shown in Figure 1-19.
2 Enter a radius for the corner round.
3 Select the Corner or Feature Select Mode in the Corner Round dialog box.
4 Select the corners or features to include.
5 Add additional corner rounds with different radii, and select corners or features for the additional corner rounds.
6 Click OK.

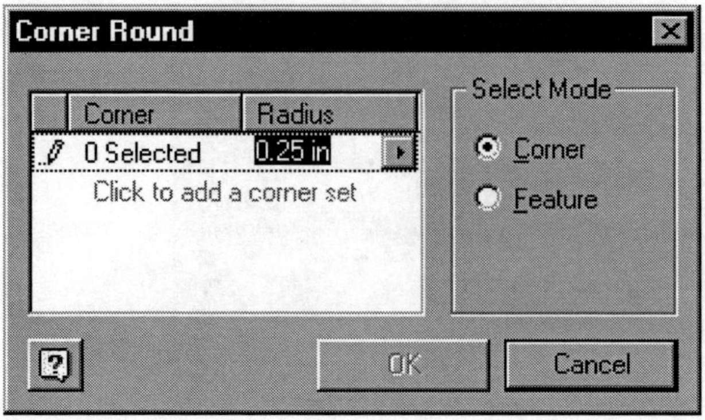

Figure 1-19

Corner Chamfer

The Corner Chamfer tool is a sheet metal specific chamfer tool. As with the Corner Round tool, all edges other than the ones at open corners of faces are filtered out.

To create a corner chamfer, follow these steps:

1. Select the Corner Chamfer tool from the Sheet Metal Panel Bar. The Corner Chamfer dialog box will appear as shown in Figure 1-20.
2. Select the chamfer style, One Distance, Distance and Angle, or Two Distances.
3. Select the corners (or edge and corner for Distance and Angle) to include.
4. Enter chamfer values.
5. Click OK.

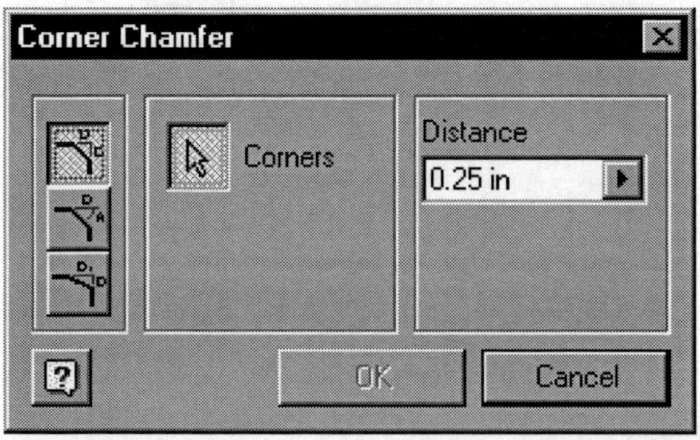

Figure 1-20

Punch Tool

Cuts and 3D deformation features such as dimples and louvers are usually added to the flat sheet metal stock in a turret punch before the sheet being bent into the folded part. Turret punch tools are positioned on the flat sheet by a center point corresponding to the tool center, and at an angle to a fixed coordinate system. The Punch tool places specially designed iFeatures on sketch points that define the center point of the tool. A streamlined interface simplifies the selection and placement of the iFeatures.

A default Punch folder is installed under the top level Catalog folder when Autodesk Inventor is installed. A selection of punch tools is included in the folder. The Punch tool lists all iFeatures in the designated punch folder. You can set the default punch folder on the iFeature tab of the Application Options dialog box.

To qualify as a PunchTool, a saved iFeature must have a single Hole Center point in the sketch of the first feature included in the iFeature. The point corresponds to the center of the tool. Nonsymmetrical shapes must have the center point position controlled by geometric relationships or equations to ensure that it remains centered when the iFeature parameters are changed. Create the iFeature in a sheet metal part to ensure that sheet metal parameters such as Thickness are saved with the iFeature.

To place a PunchTool punch, follow these steps:

1. Create a sketch on a sheet metal face. Place at least one sketch point in the sketch. Hole Center points are automatically selected as punch centers during PunchTool placement. Sketch points, line or curve endpoints, and arc centers can be manually added as additional punch centers.
2. Select the PunchTool tool from the Sheet Metal Panel Bar. The PunchTool dialog box will appear as shown in Figure 1-21.
3. Select the desired PunchTool from the list in the PunchTool dialog box. Click Next.
4. All Hole Centers are automatically selected. Hold down the SHIFT key and select Hole Centers to exclude them, and select other center points as required.
5. Select a rotation angle for all occurrences of the punch. At zero degrees rotation, the *X*-axis of the first feature in the saved iFeature is aligned with the *X*-axis of the current sketch. Click Next.
6. Enter values for the PunchTool parameters.
7. Click OK.

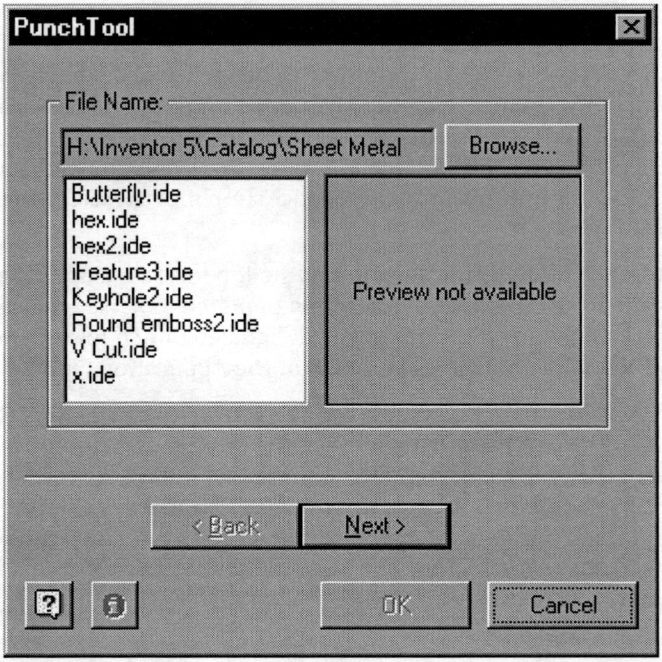

Figure 1-21

Flat Pattern

The final step is the creation of a sheet metal flat pattern that unfolds all of the sheet metal features. The flat pattern represents the starting point for the manufacturing of the sheet metal part. The flat pattern can be displayed in a 2D drawing view, complete with lines indicating bend centerlines. The Flat Pattern tool creates a 3D model of the unfolded part. You can also directly export the flat pattern in 3D (SAT file), or 2D (DXF or DWG) formats, to be used to create machine tool programming for flat pattern punching or cutting.

To create a flat pattern, follow these steps:

1 Select a face. The selected face will remain fixed and all other faces will unfold from this face.

NOTE This is not a strict requirement, but selecting a face before creating a flat pattern is good practice.

---

2 Select the Flat Pattern tool from the Sheet Metal Panel Bar. The flat pattern model is displayed in a new window. The flat pattern model is saved in the same document as the folded model. A Flat Pattern node is also added to the Browser as shown in Figure 1-22.

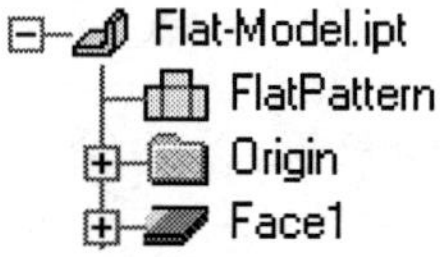

Figure 1-22

3 Right click Flat Pattern at the top of the Browser. Menu items are available for saving the flat pattern, aligning the flat pattern to a model edge, and displaying the overall dimensions (Extents) of the flat pattern.

4 To document a flat pattern start a new drawing. Place a drawing view and select Flat Pattern from the Presentation View list in the Create View dialog box.

Common Tools

A number of tools on the Sheet Metal Panel Bar are common to sheet metal and standard parts. The following are short descriptions of how to use these tools in the sheet metal environment:

Work Features — Work planes are often used to define sketch planes for disjointed faces.

Holes — The Hole feature is common to both the part modeling and sheet metal environments. You can enter sheet metal parameters such as Thickness in numeric fields such as hole depth.

Catalog Tools — iFeatures can be saved and placed in the sheet metal environment. See the PunchTool section for information on special iFeatures for sheet metal.

Mirror and Feature Patterns — Most sheet metal features cannot be mirrored or patterned. Holes, cuts, iFeatures, and features created with the standard part feature tools can be patterned or mirrored.

Promote — IGES files can be imported in the sheet metal environment.

Detailing Sheet Metal Designs

Drawing views of both the folded 3D model and flat pattern can be created to detail a sheet metal part. The view of the 3D model details the folded shape of the part. The flat pattern provides the ability to annotate all bend locations, relief sizes and cutouts using the tools available on the Drawing Annotation Panel Bar in Autodesk Inventor.

A flat pattern must be generated in the 3D model in order for the flat pattern to be placed on a 2D drawing sheet. If the flat pattern is deleted in the model after the creation of a flat pattern drawing view, the drawing view will also be deleted and you will not be able to create a drawing view of the flat pattern.

Drawing views are created using the Create View tool when working in Autodesk Inventor's drawing environment. Select either the Folded Model or Flat Pattern options available on the Presentation View drop list to place the appropriate view. You can then activate the Drawing Annotation Panel Bar and use all of the annotation tools to detail the drawing views.

NOTE Activate the Create View tool two times to place a view of both the folded model and the flat pattern on a drawing sheet. Select the Folded Model option from the Presentation View drop list, and then select the Flat Pattern option when activating the tool the second time.

Exercise 1: Sheet Metal in Assemblies

Sheet metal brackets, guards, and enclosures often rely on the size and position of other assembly components. You can design sheet metal parts in-place in an assembly using similar techniques used to create standard parts in an assembly. Cross part projected edges and loops can be used to define face edges. Most sketch-based sheet metal features can be made adaptive, but some sheet metal parameters such as the bend angle of a flange cannot adapt to meet assembly constraints. In addition to adaptive sheet metal design, you can create associative sheet metal parts in assemblies using the master model technique examined later in this course.

Cross Part Associativity

You will project edges from the existing components, and use those edges to create sheet metal features. In order to maintain an associative relationship to the original edges, you must have cross part associativity activated in Autodesk Inventor.

1 Select Tools > Application Options from the main menu.
2 Click on the Assembly tab.
3 Ensure that Enable Associative Edge/Loop Geometry Projection During In-Place Modeling is checked as shown in Figure 1-23.
4 Click OK.

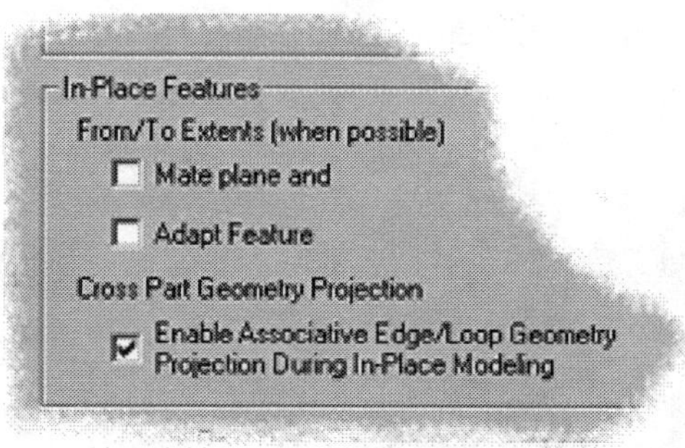

Figure 1-23

In-Place Part Creation

In this exercise, you open an existing assembly and create a sheet metal bracket to support the two parts in the assembly using various sheet metal tools.

1 Open the *SM-Assembly.iam* file. The assembly should match the one shown in Figure 1-24.

Figure 1-24

2 From the Assembly Panel Bar, click the Create Component tool.

3 In the Create In-Place Component dialog box:

   a. Click the Browser button next to Template and select the *Sheet Metal (in).ipt* template from the English tab.

   b. Click OK to accept the template.

   c. Enter **SM03** in the New File Name edit box.

   d. Ensure that Constrain sketch plane to face or plane is checked.

   e. Click OK.

4 Click on the face highlighted in Figure 1-25.

Figure 1-25

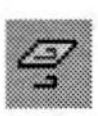

5 From the Sketch Panel Bar, click the Project Geometry tool. Click on the three edges of the large box highlighted in Figure 1-26.

TIP Avoid clicking the face of the box; the loop around the face will be projected if you do.

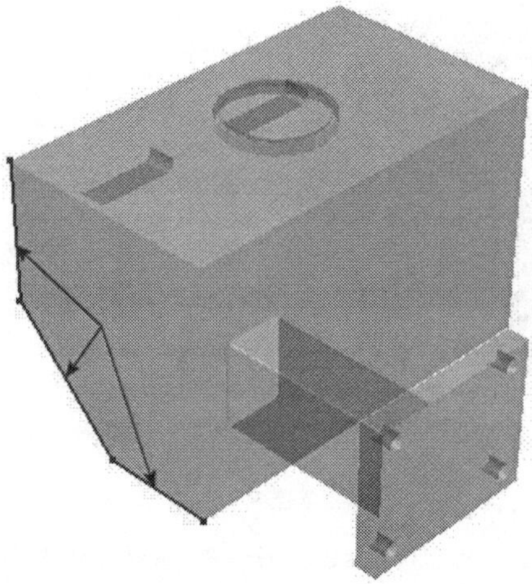

Figure 1-26

6 Click the Return tool on the Standard Toolbar to return to the Sheet Metal environment.

7 From the Sheet Metal Panel Bar, click the Contour Flange tool.

8 Click one of the three projected edges.

9 In the Contour Flange dialog box:

   a. Click the Offset tool.

NOTE The preview of the contour flange now shows the material offset to the outside of the box.

   b. Enter **14** in the Distance edit box.

   c. Click the Flip Direction button so the contour flange will extrude along the length of the box.

   d. Click OK.

Your view should resemble Figure 1-27.

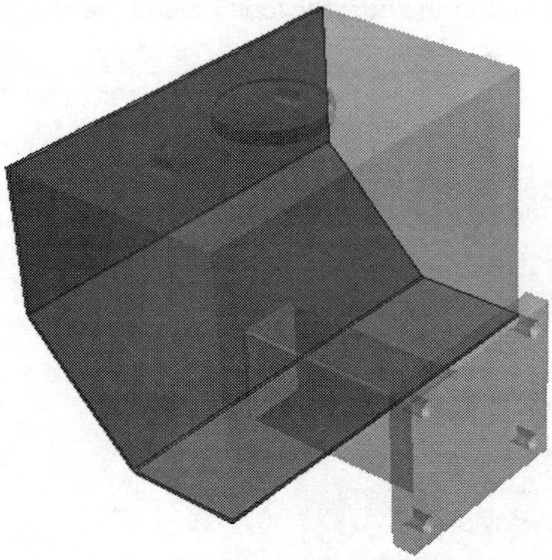

Figure 1-27

The contour flange is marked as adaptive in the Browser, and the part is marked adaptive. The sketch consumed by the contour flange has three references as children, one for each edge projected into the sketch. The contour flange matches any changes to these edges in the box.

Add Sheet Metal Faces

1 From the Standard Toolbar, click the 2D Sketch tool.
2 Click on the face highlighted in Figure 1-28.

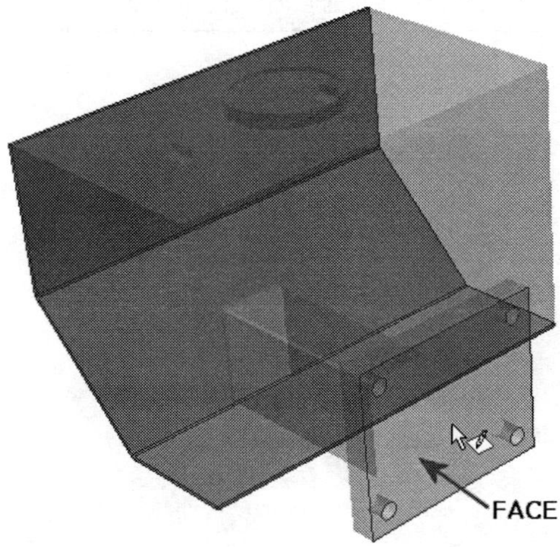

Figure 1-28

An adaptive work plane is created in the sheet metal part. A zero-offset flush constraint positions the work plane on the face. A new sketch is activated on the work plane.

3 From the Sketch Panel Bar, click the Project Geometry tool.
4 Click the left vertical edge of the flange as shown in Figure 1-29.

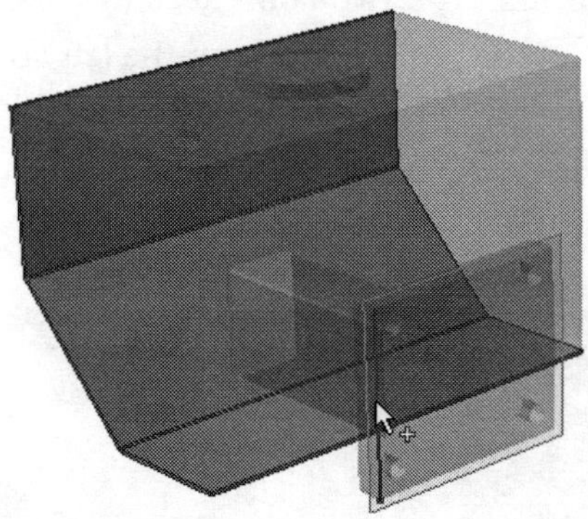

Figure 1-29

An associative reference to the projected edge is generated, and shown in the browser under the current sketch.

5 With the Project Geometry tool still active, click the edge of the contour flange as shown in Figure 1-30.

TIP You can project endpoints as well as edge geometry.

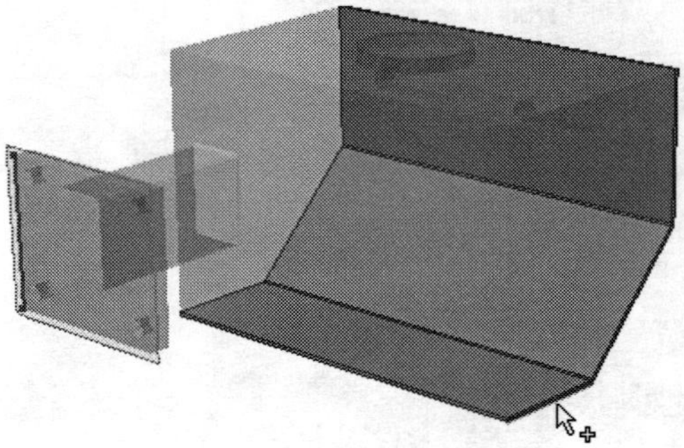

Figure 1-30

Since the projected edge is from the same part, no cross part reference is required.

TIP Keep cross part associations to a minimum. The fewer dependencies you have between components, the less chance for errors to occur during subsequent edits.

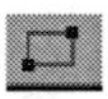

6 From the Sketch Panel Bar, click the Two Point Rectangle tool.

7 Click the upper endpoint of the projected line on the left and the point projected from the edge of the contour flange. Your sketch should match the one shown in Figure 1-31.

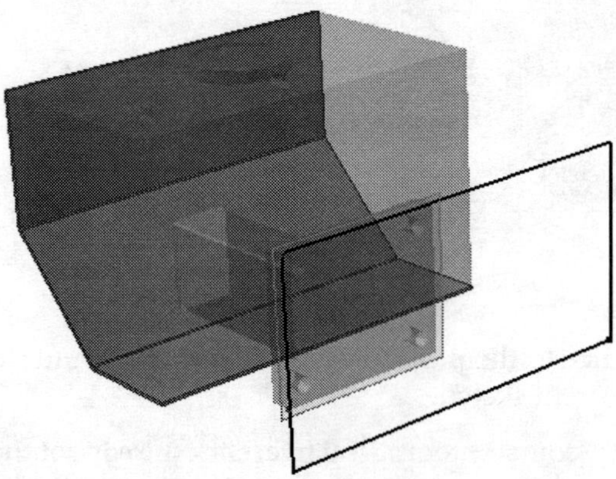

Figure 1-31

8 Finish the sketch.

9 Click the Face tool on the Sheet Metal Panel Bar.

10 In the Face dialog box:

a. Click the Select Edge tool.

b. Click the Edge highlighted in Figure 1-32.

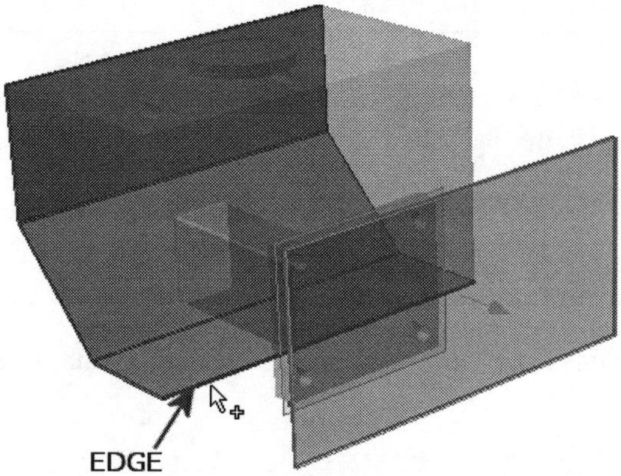

Figure 1-32

11 Click OK.

Your part should match the one shown in Figure 1-33.

Figure 1-33

You can select an existing edge during the creation of a face feature. The selected edge must be parallel to the adjacent edge on the new face. One or both faces are extended or shortened, and a bend is added at the intersection between the two faces.

NOTE This sheet metal face could easily have been incorporated into the original contour flange. The projected edge of the box support is connected to the lower projected edge with two sketch line segments. We split the part into separate features for demonstration purposes.

---

12 Start a new sketch on the underside of the cylindrical part (*AssyComp02.ipt*).

13 From the Sketch Panel Bar, click the Project Geometry tool.

14 Click on the sketch face to project all loops on the face into the sketch.

TIP Do not click individual edges on the face.

A single reference under the current sketch contains the adaptive link to all loops on the face.

15 Finish the sketch.

16 Create a sheet metal face and bend that connects to the edge of the previous face.

When complete, the sheet metal part should resemble Figure 1-34.

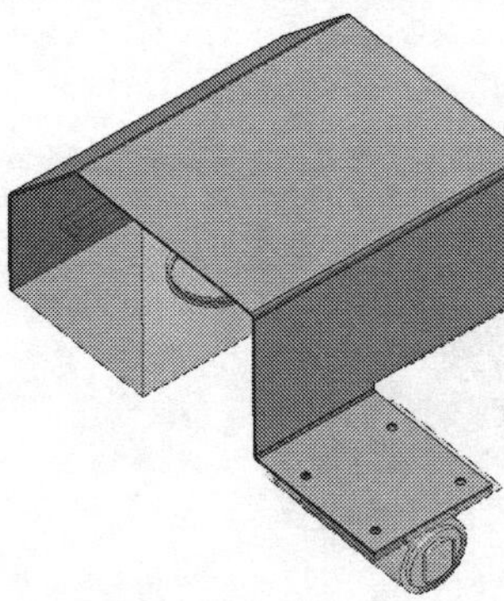

Figure 1-34

Adding Sheet Metal Features

The key faces of the sheet metal part are in place. You now add flanges, corner seams, and a cut feature to complete the design.

1 Clean up the view by turning off the visibility of other parts in the assembly and the work planes in the sheet metal part.

2 Reorient your view to resemble Figure 1-35.

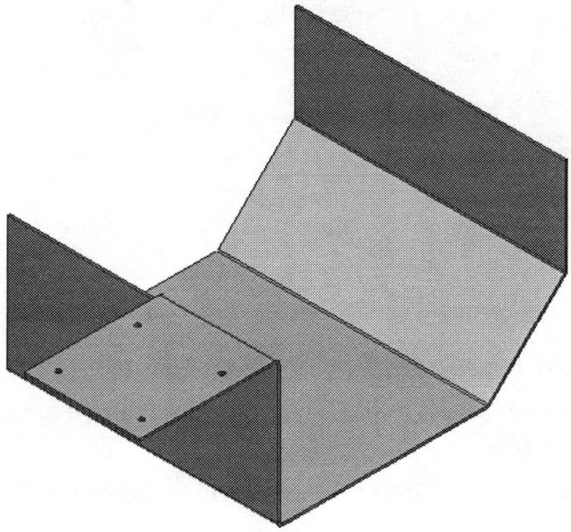

Figure 1-35

3 From the Sheet Metal Panel Bar, click the Flange tool.

4 Click the edge highlighted in Figure 1-36.

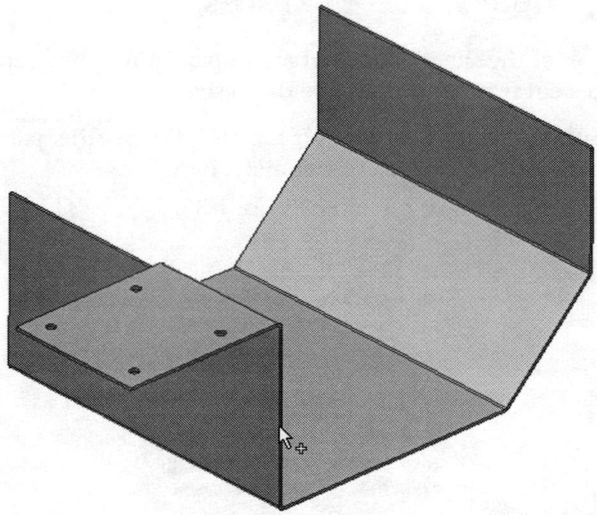

Figure 1-36

5 In the Flange dialog box:

a. Click the More button in the lower right corner to expand the dialog box.

b. Select Offset from the Extents Type list.

c. Click the lower endpoint of the edge and then the upper endpoint of the edge.

d. Enter 0 in the Offset1 edit box.

e. Enter 2 in the Offset2 edit box.

f. Click Apply.

Your part should match the one shown in Figure 1-37.

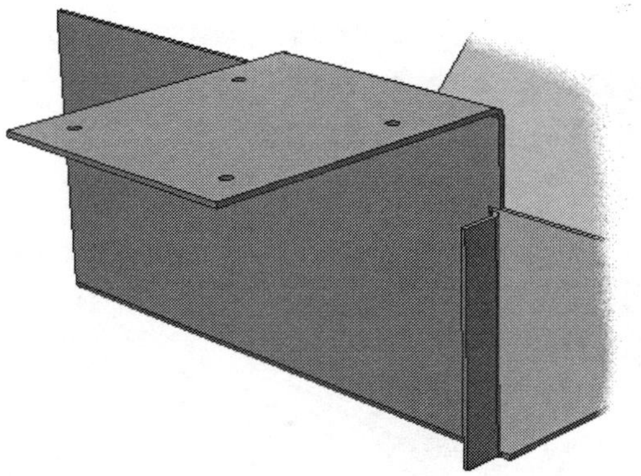

Figure 1-37

The Flange tool remains active.

6 Click the edge highlighted in Figure 1-38.

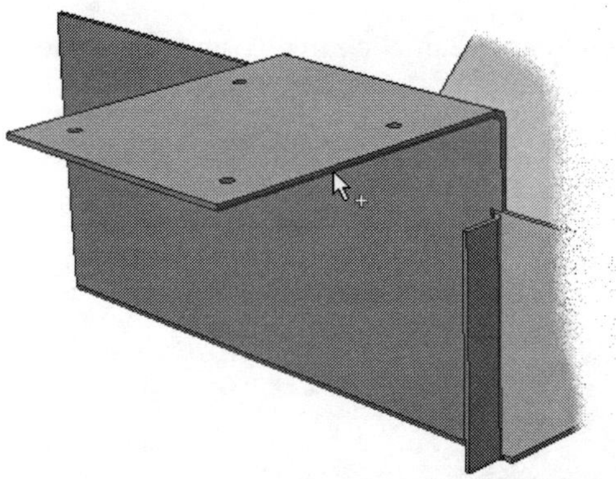

Figure 1-38

7 In the Flange dialog box:

  a. Click the left endpoint of the edge.

  b. Click the right endpoint of the edge.

  c. Click OK.

The second flange should match the one in Figure 1-39.

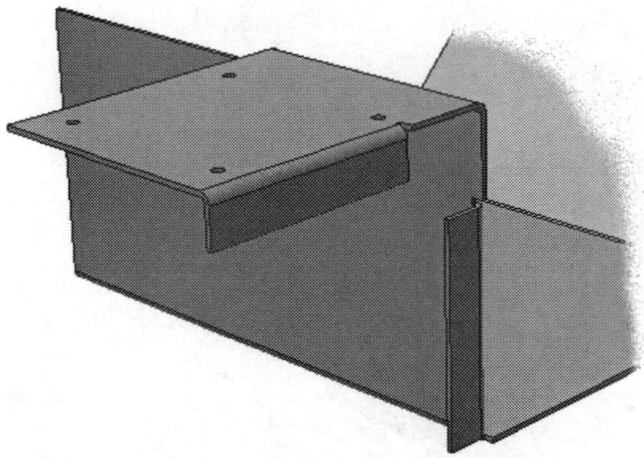

Figure 1-39

Add Corner Seams

The flanges are set back from the corner to enable a mitered corner seam to be applied.

1 From the Sheet Metal Panel Bar, click the Corner Seam tool.
2 Click the two edges highlighted in Figure 1-40.

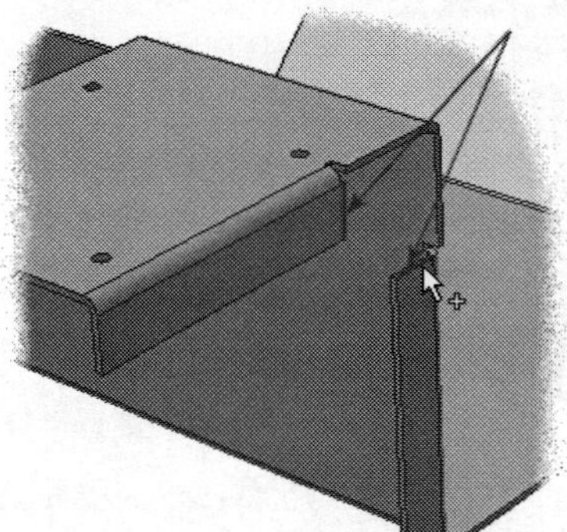

Figure 1-40

3 In the Corner Seam dialog box:
   a. Enter **Thickness/2** in the Gap edit box.
   b. Click OK to create the Flange shown in Figure 1-41.

Next, you create a cut that crosses one of the sheet metal bends. During manufacturing, the cut is made in the blank sheet, and the bend is then added in a press brake. You can unfold the sheet metal faces into the sketch for the cut and then locate the cut feature on the flattened sheet. The sheet metal Cut tool enables the cut to properly form around the bend in the folded state.

4 Create a new sketch on the face highlighted in Figure 1-41.

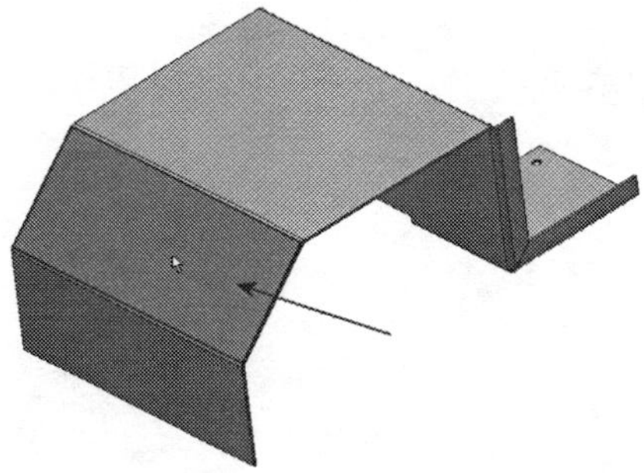

Figure 1-41

5 In the Sketch Panel Bar, click the arrow next to the Project Geometry tool, and then click the Project Flat Pattern tool.

6 Click the face highlighted in Figure 1-42.

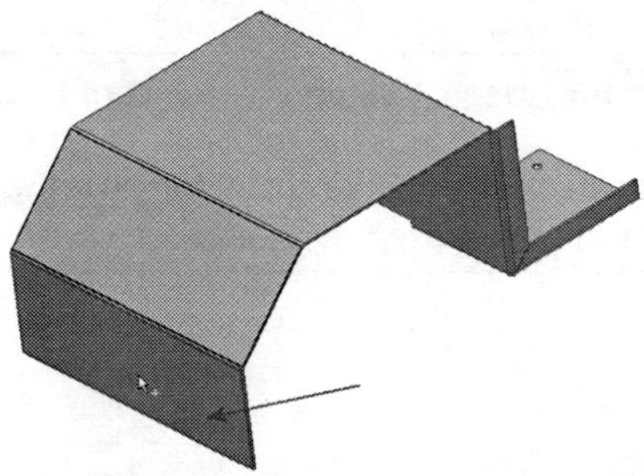

Figure 1-42

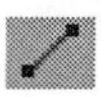

7 Use the Line tool to sketch a profile similar to the one in Figure 1-43. Ensure the sketch edges cross the bend zone at an angle.

NOTE You can add dimensions to the sketch, but they are not required.

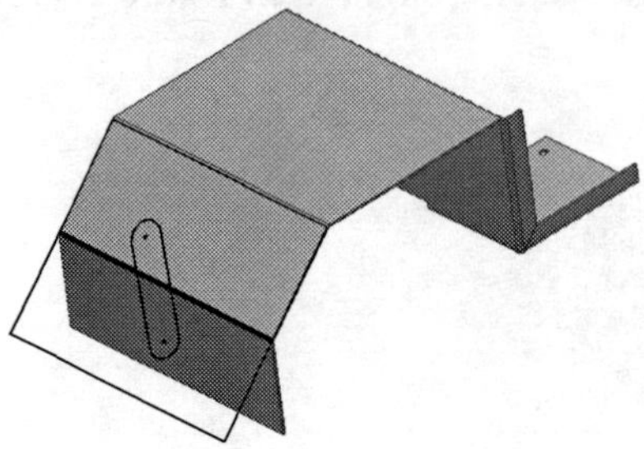

Figure 1-43

8 Finish the sketch.

9 From the Sheet Metal Panel Bar, click the Cut tool.

10 Click inside the obround shape.

11 In the Cut dialog box, click the Cut Across Bend option.

12 Click OK to complete the cut shown in Figure 1-44.

NOTE If the cut fails, turn off the Adaptive setting for the contour flange (first feature in the part). Right-click Contour Flange1 in the Browser, and select Adaptive to clear the check mark. After completing the cut feature, use the same technique to restore adaptivity for this feature.

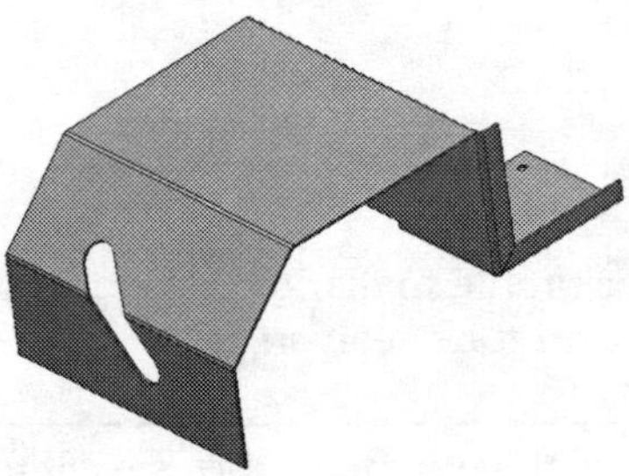

Figure 1-44

Flat Pattern Creation

You can create a flat pattern model of a sheet metal part at any time in the design process. As you add features to the folded part, the flat pattern model is automatically updated. If a new feature would cause the part to become unfoldable, an error message is displayed.

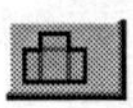

1 From the Sheet Metal Panel Bar, click the Flat Pattern tool.

The flat pattern model is displayed in a new window as shown in Figure 1-45.

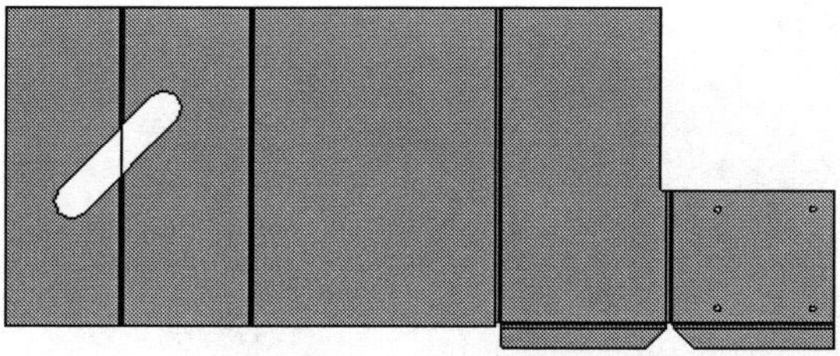

Figure 1-45

You can place both the folded and unfolded versions of the model in drawing views. A flat pattern drawing can be annotated within Autodesk Inventor, or you can export the drawing in DWG or DXF format. You can also export as a solid model in SAT format, directly from the flat pattern model.

2. Right-click on Flat Pattern in the Browser.
3. Select Save Copy As from the menu.
4. Select DXF Files (*.dxf) from the Save as type list in the Save dialog box.
5. Enter Flat Pattern for the filename.
6. Click the Options button.
7. In the DXF File Export Options dialog box:
   a. Select the R12 File Version.
   b. Click Next.
   c. Examine the Style to Layer mapping, and then click Finish to close the wizard.
   d. Click Save.

NOTE You can change the names associated with the Layer Mapping as desired.

NOTE Volo View 2 or Volo View Express 2 needs to be installed for DWG or DXF export.

---

8. Close the Flat Pattern window.
9. Click the Return tool on the Standard Toolbar to return to the assembly level.
10. Turn on the visibility of the other two parts in the assembly.

The assembly should resemble Figure 1-46.

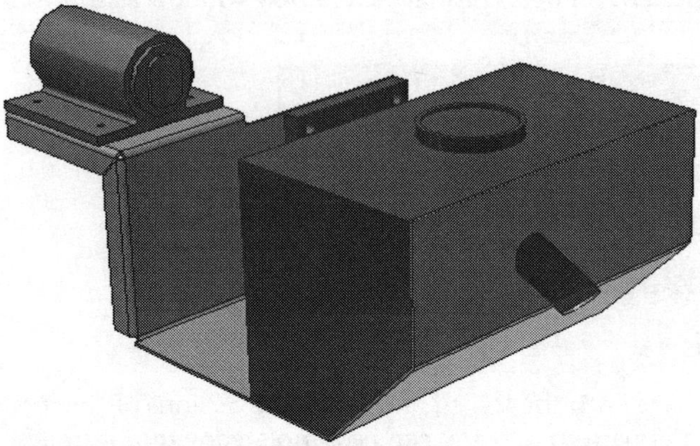

Figure 1-46

Adaptive Sheet Metal Update

Next, you make some changes to the assembly, and the sheet metal part adapts to match the revisions.

1 Expand *AssyComp01.ipt* in the Browser.
2 Click on the Mate (–1.500 in) assembly constraint node.
3 In the Offset edit box at the bottom of the Browser, enter **3 in.** as the new offset.

The position of the cylindrical part is updated and the sheet metal part updates to match as shown in Figure 1-47.

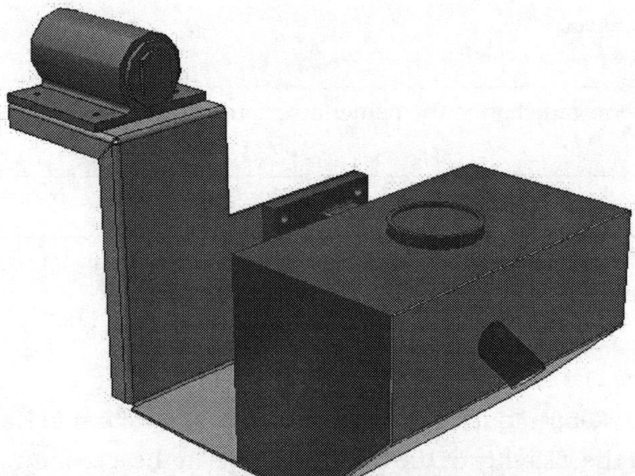

Figure 1-47

4 Right-click *AssyComp01.ipt* in the Browser.

5 Select Edit from the menu.

6 Right-click Extrusion1 under *AssyComp01.ipt*.

7 Select Edit Sketch from the menu.

8 Edit the dimensions shown to match those shown in Figure 1-48.

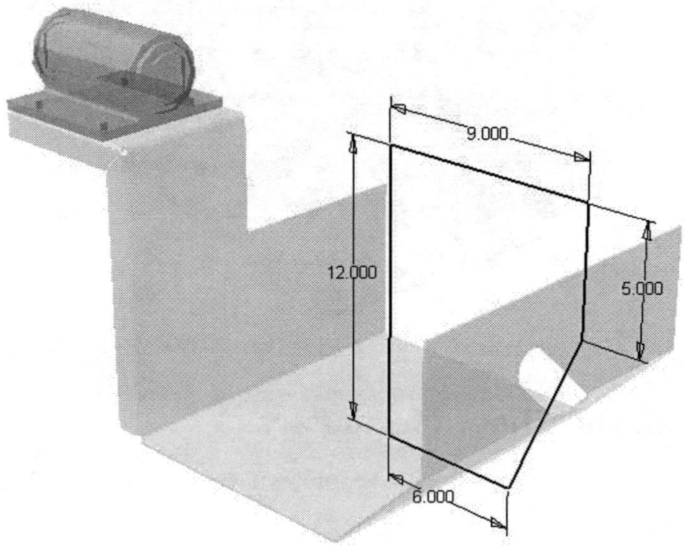

Figure 1-48

9 Return to the assembly.

The sheet metal part adapts to match the new shape of the box as shown in Figure 1-49.

TIP It is good practice to turn off adaptivity once you have completed any edits that would cause the part to adapt. You can turn adaptivity back on if further edits are required later in the design process.

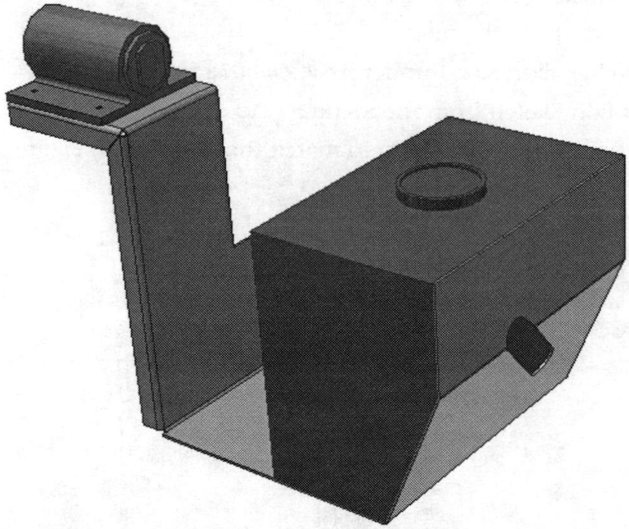

Figure 1-49

Detail the Sheet Metal Part

Finally, you create a drawing that contains a view of both the 3D model and the flat pattern.

1 Start a new drawing based on the ANSI (mm).idw template from the Metric tab of the Templates dialog box.

2 From the Drawing Management Panel Bar, select the Create View tool.

3 Click the Explore Directories button next to the File box.

4 Select the *SM03.ipt* file in the Open dialog box.

NOTE You created this sheet metal part in this exercise.

---

5 Click Open.

6 In the Create View dialog box:

a. Select Flat Pattern from the Presentation View drop list.

b. Specify a scale of 0.25.

c. Select Pivot Right from the View list.

d. Choose Hidden Line for the Style.

e. Click a location near the center of the drawing to place the view.

Add a Folded Model View

1 From the Drawing Management Panel Bar, select the Create View tool.
2 Click the Explore Directories button next to the File box.
3 Select the *SM03.ipt* file in the Open dialog box.
4 Click Open.
5 In the Create View dialog box, select Folded Model from the Presentation View drop list.

6 Select the Change View Orientation button.
7 Use the View tools to orient the model so you can see all sides of the part.

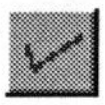

8 On the Custom-View Toolbar, select the Exit Custom View button.
9 In the Create View dialog box:
   a. Specify a scale of 0.25.
   b. Select the Shaded View style.
   c. Click a location on the drawing sheet to place the view.

Your drawing should resemble Figure 1-50.

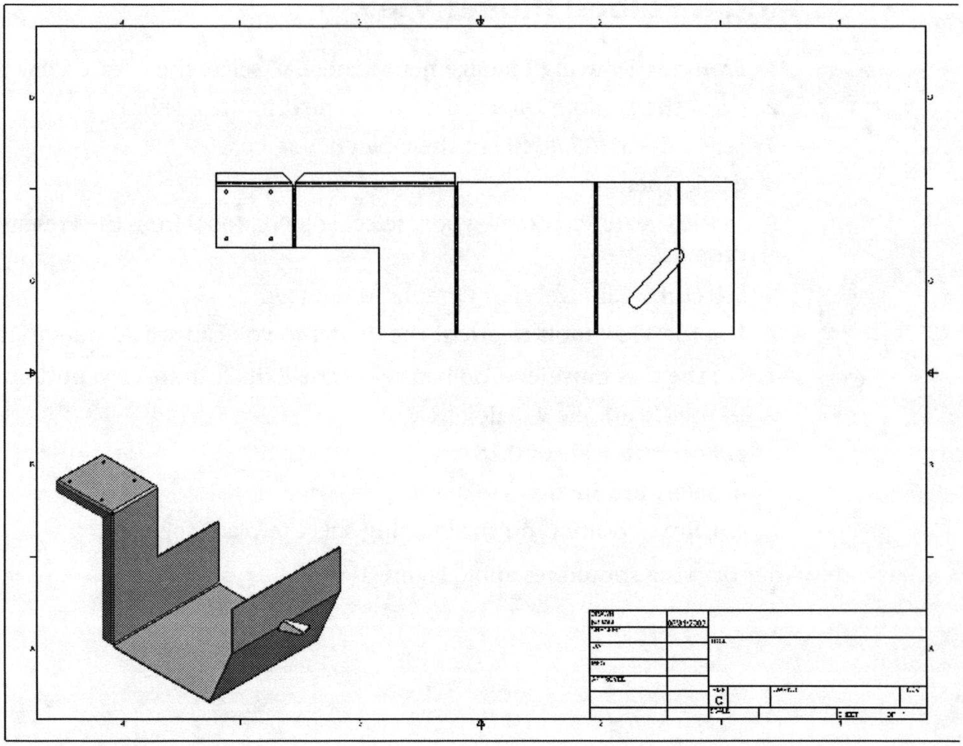

Figure 1-50

10 Click the title of the Drawing Management Panel Bar and select Drawing Annotation from the menu.

11 Select the General Dimension tool.

12 Create some dimensions to detail the flat pattern view.

13 Close the files without saving.

End of exercise.

Face and Flange Techniques

In most sheet metal designs, the folded shape of the part is known. You can create the folded model using various techniques, the most basic being a process of adding individual faces, flanges, and other sheet metal features to build the final state of the folded part. A key face is the first feature added to the part, and additional "as bent" features are added, and joined to open edges of the part. During feature creation, a bend is automatically created at the intersecting edge of the new feature and the existing part. You can also add bends between disjointed faces, a technique that is very helpful when designing a sheet metal component in the context of an assembly. When the folded design is complete, Autodesk Inventor can create a flattened version of the model, commonly called a flat pattern.

The flat pattern locates the position of bend centerlines used to form the part in a press brake or other sheet metal forming tools. The flat pattern also contains any holes, cutout, and other features placed on faces of the folded part. The size of the flat pattern is determined by the settings in the current sheet metal style.

When a metal blank or sheet is bent in a press brake, the metal in the bend area deforms. Material on the inside of the bend compresses, and material on the outside of the bend is stretched. This deformation must be taken into account when calculating the flattened state of the folded model. The amount of deformation is dependent on the material, the radius and angle of the bend, and the process and equipment used to create the bend.

Within the sheet there is a plane where the material neither compresses nor stretches. If the location of this plane is known, the length of the flat sheet can be calculated. For many materials and processes, this location of this neutral plane is known, and can be expressed as a percentage of the thickness of the part, as measured from the surface on the inside of a bend. This offset value is referred to as a KFactor. Figure 1-51 shows the neutral plane of a sheet metal bend. The location of the neutral plane is KFactor * Thickness, where 0° < °KFactor < 1. The default sheet metal style uses a KFactor of 0.44. This value is appropriate for many common materials and processes.

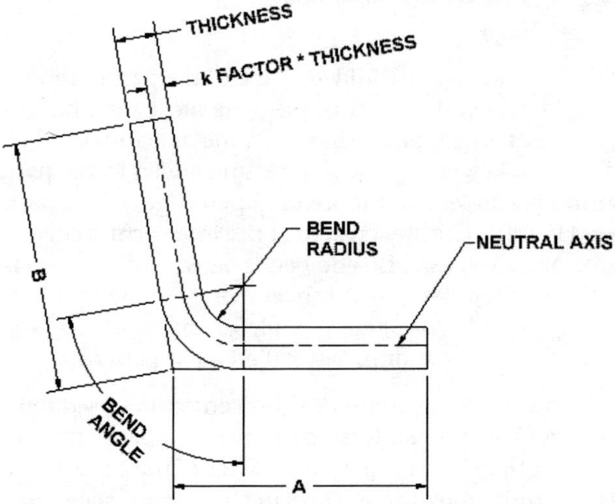

Figure 1-51

You can edit the default sheet metal style, or add additional styles. You store sheet metal styles in a template, and apply the appropriate style for the design. The current sheet metal style applies to all sheet metal features in the part. Although you cannot apply separate styles to different sheet metal features in a part, you can override default values during feature creation or edit.

For example, some materials have different bending properties dependant on the orientation of the bend relative to the grain of the material. As the sheet is manufactured, the material microstructure aligns with the rolling direction. This results in a sheet with bend properties related to its orientation in the press brake. Stainless steel sheets often display this anisotropic behavior.

The use of a constant KFactor is not always appropriate. For some materials and processes, or when formed parts require precise tolerances, a bend table can be used in place of the KFactor. A bend table uses empirically derived information to apply length adjustments to bends. With overrides, you can apply a separate KFactor or bend table to each bend on a sheet metal part.

Bend Tables

In the calculation of the unfolded length of a bend, a bend table replaces the KFactor with a set of known adjustments to the unfolded length. A bend table can provide greater precision of unfolded dimensions since the values are derived from your measurements of bend length of a particular material on a specific machine.

When a bend table is used to calculate an unfolded length, the following formula applies:

L=A+B-x

Referring to Figure 1-52, the variables in the formula are

L	Unfolded Length
A	Length of folded face 1
B	Length of folded face 2
x	Adjustment from bend table

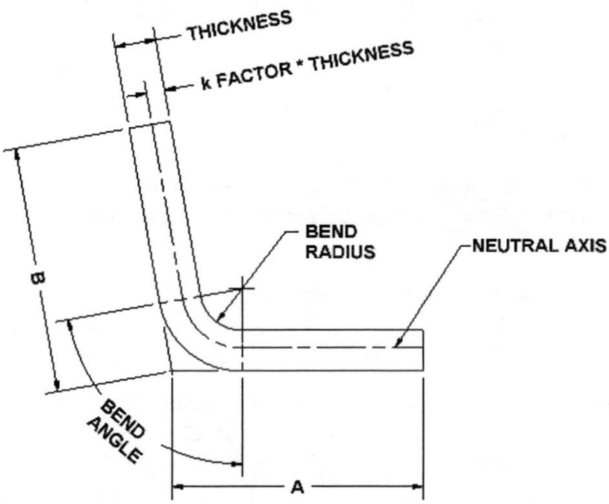

Figure 1-52

The value of x is dependent on sheet thickness, angle of the bend, and the inner radius of the bend.

NOTE The measurements A and B are to the intersection of the extended outer faces on either side of the bend. This intersection is used when the angle of the bend is less than or equal to 90°. Figure 1-53 is used when the bend angle is greater than 90°. The measurements are parallel to the face, and tangent to the outer surface of the bend. The same formula is used to determine the unfolded length of the part.

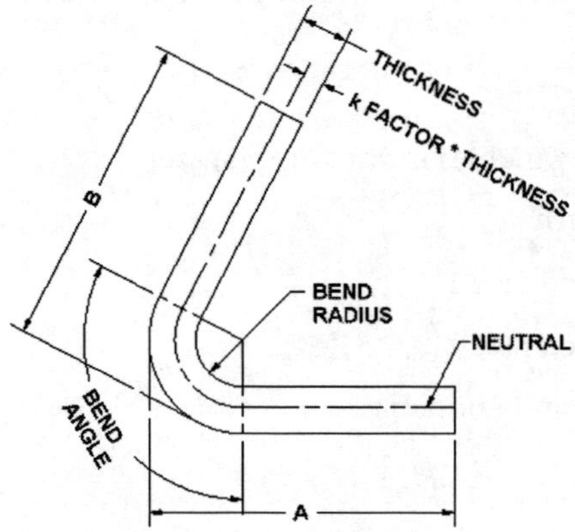

Figure 1-53

Bend Table Editing

The bend table must be a plain text file (.txt extension), but you can create or edit a spreadsheet version of the bend table using Microsoft Excel, and then save the table in text format.

Autodesk Inventor ships with both metric (mm) and English units (in.) bend tables that you can modify. Each bend table is supplied in both .txt and .xls formats. The following exercise includes the setup of a custom bend table.

Figure 1-54 illustrates the angle information required in a bend table. A flat sheet bent at an angle of 110° will have an opening angle of 70°. The opening angles are entered in the bend table.

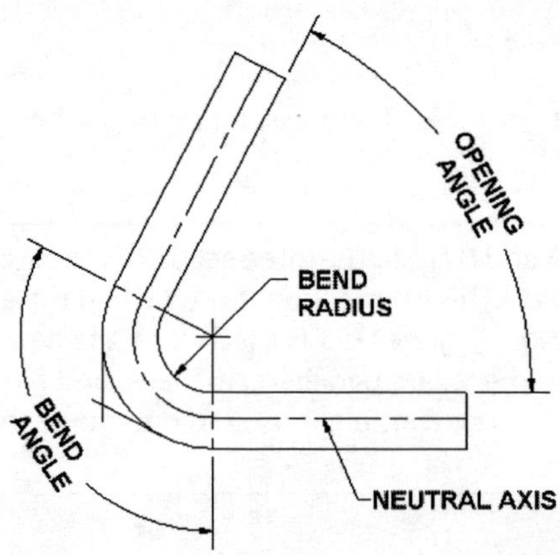

Figure 1-54

Bend and Corner Reliefs

A sheet metal style also determines the treatment of bend reliefs. Bend reliefs can be specified when a bend zone (the area deformed during a bend) does not completely cross a face (see Figure 1-55).

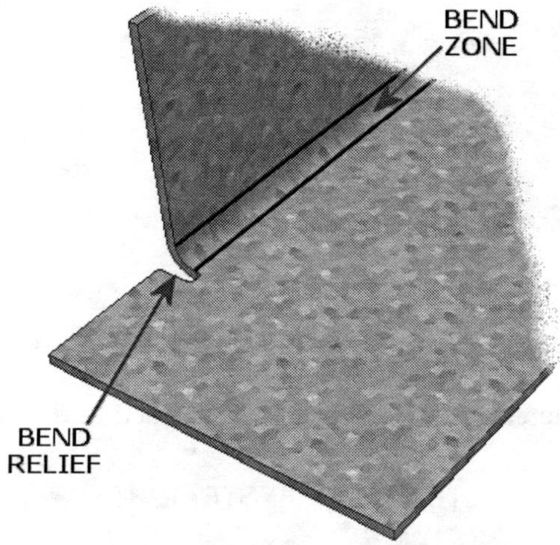

Figure 1-55

If bend reliefs are used, they are incorporated in the flat blank before folding. You can generate bend reliefs by punching, or with laser, water-jet, or other cutting methods. The creation of bend reliefs increases production costs. It is common practice with thin, deformable materials such as mild steel to add bends without bend reliefs. The material is allowed to tear where the bend zone interests with the adjacent face (see Figure 1-56). You cannot specify None as the default bend relief, so you must override the default bend relief style during feature creation. During the creation of any feature that generates a bend, you can uncheck the bend relief option to produce a torn bend relief.

Figure 1-56

Figure 1-57 illustrates the bend relief types available on the Bend tab of the Sheet Metal Styles dialog box. When a bend relief is close to an open edge, the Minimum Remnant setting determines if the relief is extended to the face edge.

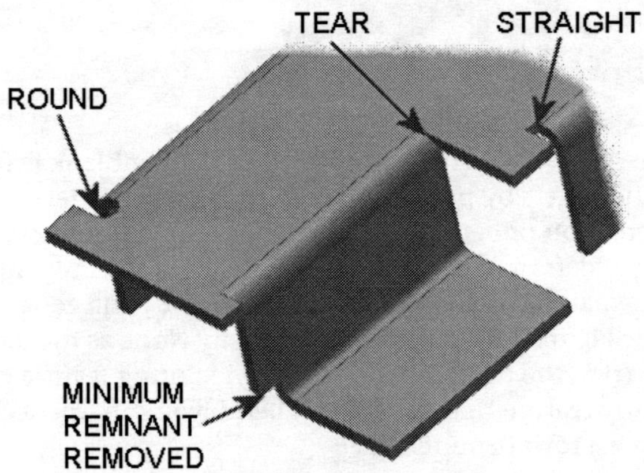

Figure 1-57

When three faces intersect, a corner seam feature controls the gap between the open faces, and the relief shape at the intersection. As with bend reliefs, corner reliefs are added to the flat sheet before bending. Unlike bend reliefs, you can specify ify a default Tear corner seam. You work with corner reliefs in Exercise 7: Shelled Solids.

Exercise 2: Creating Faces and Flanges

In this exercise, you use the basic building blocks of sheet metal parts, faces, and flanges, to create a simple bracket.

The design requires precision forming. You create a bend table from empirical data gathered from experience with the tooling and material used to manufacture the part.

The bend table and sheet metal style enable you to unfold the bent part to an accurate flat pattern model. The completed exercise is shown in Figure 1-58.

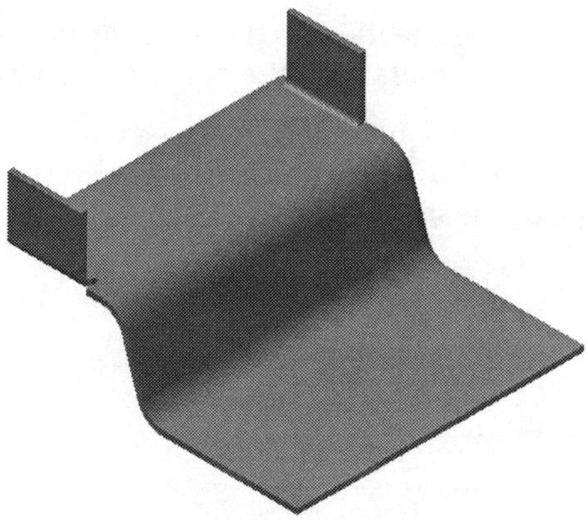

Figure 1-58

The bracket has two large radius bends and two smaller radius bends that are at right angles to the larger bends. The bracket is made from stainless steel.

To prevent cracks from forming in the bend zones, the part requires different bend radii for bends with and against the grain of the material. In addition, the large radius bends are formed on a machine that requires a bend table for accurate bend length calculation.

Creating a Bend Table

1 Open the *SS Table.xls* file located in the folder where you installed the training materials.

NOTE The default installation directory is ...\Program Files \Autodesk \Mastering Sheet Metal Design.

NOTE This spreadsheet is a smaller version of the *Bend Table (mm).xls* file included in the Samples\Bend Tables folder under your Autodesk Inventor installation folder.

2 Scroll down to the bottom of the spreadsheet. The correction values for different bend angles and bend radii are shown in cells with a gray background.

3 Click in cell C72 (Opening Angle = 45°, Bend Radius = 6 mm).

TIP The complex equation for the cell is displayed in the Formula Bar at the top of the screen.

You can adjust this formula to obtain the measured bend adjustment value, but it is far more practical to replace the formula with the empirical values.

Most of the calculated values in the table have been replaced with static values. The table in Figure 1-59 outlines the measurements obtained from test bends of the material where the open angle of the bend is 45° (bend angle = 135°).

Unfolded Length (L)	Open Angle	Bend Radius	Leg 1 (A)	Leg 2 (B)	Adjust (x)
200.00	45.0	6.00	110.55	90.40	**0.95**
200.00	45.0	9.00	112.40	87.25	**-0.35**
200.00	45.0	12.00	109.85	88.45	**-1.70**

Figure 1-59

4 Enter the three values from the last column of the above table in the corresponding cells in the bend table (see Figure 1-60).

64	*** TABLE 1				
65	;				
66	;sheet thickness				
67	/S	3.040000			
68	;				
69	;bending radii				
70	/R		6.000000	9.000000	12.000000
71	;opening angle:	----------	---------- correction value x	---------	
72	/A	45.000000	0.950000	-0.350000	-1.700000
73	/A	60.000000	2.900000	2.350000	1.900000
74	/A	75.000000	4.750000	4.950000	5.200000
75	/A	90.000000	6.600000	7.700000	8.850000
76	/A	105.000000	4.300000	4.750000	5.350000
[illegible]	/A	120.000000	[illegible]	[illegible]	[illegible]

Figure 1-60

NOTE If the bend angle or bend radius falls between the table values, a linear interpolation is used to calculate the bend correction (x).

---

5 From the Excel main menu, select File > Save As.

6 Select Text (Tab delimited) (*.txt) from the Save As type list.

7 Enter **SS-3mm-With Grain.txt** in the File name edit box. You must save the bend table as a text file to use it in a sheet metal style.

8 Click Save and close the worksheet without saving changes.

Sheet Metal Style Setup

You now set up a sheet metal style with two unfold solutions. A KFactor is used for bends against the material grain, and the above bend table is used for bends parallel to the grain.

1 Start a new sheet metal part based on the *Sheet Metal (mm).ipt* template from the Metric tab of the Templates dialog box.

2 From the Sketch Panel Bar, click the Two Point Rectangle tool.

3 Sketch and dimension a rectangle to match the one in Figure 1-61.

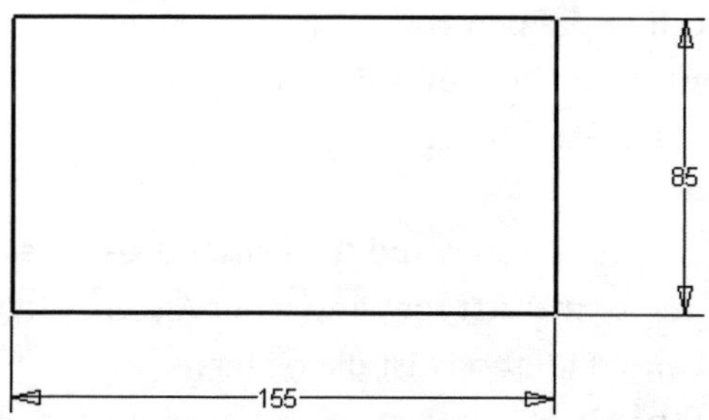

Figure 1-61

4 Finish the sketch.

Update the Sheet Metal Style

1 From the Sheet Metal Panel Bar, click the Styles tool.

2 In the Sheet Metal Styles dialog box:

a. Click New.

b. Enter **SS-3mm** in the Style List edit box.

c. Select Stainless Steel, 440C from the Material list.

d. Enter **3.04 mm** in the Thickness edit box.

e. Click Modify List to add new KFactors and bend tables.

3 In the Unfold Method List dialog box:

a. Click New.

b. Click in the Name cell of the new row (KFactor1), and enter **With Grain** in the cell.

c. Click in Type cell of the new row, and then select Bend Table from the list as shown in Figure 1-62.

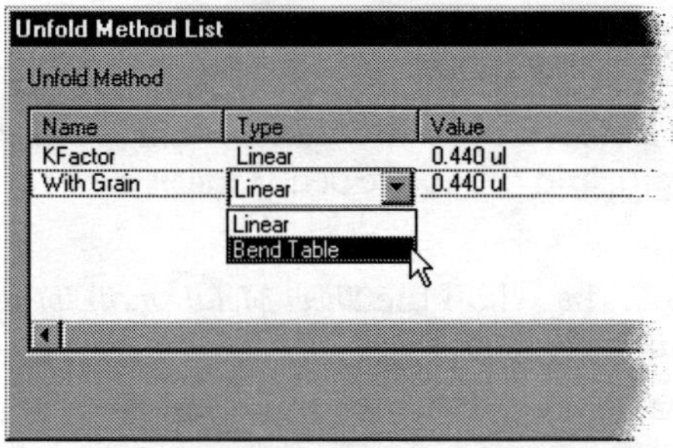

Figure 1-62

4 Click in the Value cell of the new row.

5 Click the ellipses button at the right side of the cell.

6 Select the *SS-3mm-With Grain.txt* file.

7 Click Open.

8 Click the Value cell in the first row, and then enter 0.465 ul as the new KFactor.

9 Click in the Name cell of the first row, and enter Against Grain.

10 Click Done in the Unfold Method List dialog box.

11 On the Bend tab of the Sheet Metal styles dialog box:

a. Select Arc from the Transition list.

b. Select Round from the Relief Shape list.

c. Enter **3 mm** in the Radius edit box.

d. Enter **Thickness * 4 ul** in the Minimum Remnant edit box.

12 Click Save.

13 Select SS-3mm from the Active Style list.

14 Click Done.

Creating Faces and Bends

You first create a face from the existing sketch. Next, you create two flanges that join onto the existing part with bends that are parallel to the grain of the sheet.

The default bend radius is overridden to match one of the bend radii specified in the bend table, and the bend table is used to calculate the unfolded length of the bend.

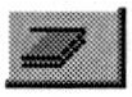

1 From the Sheet Metal Panel Bar, click the Face tool.

You can change the extrusion distance for the first face in a part, but the face is always extruded a distance equal to the Thickness value in the current sheet metal style.

2 Click OK.

3 Change to an Isometric View of the part.

NOTE A part based on a sheet metal template automatically adds a number of model parameters to the part. Thickness, BendRadius, and other values are added to the model parameters list. A standard Autodesk Inventor part also adds these model parameters if the sheet metal application is selected from Tools > Applications on the main menu.

4 From the Sheet Metal Panel Bar, click the Flange tool.

5 Click the edge highlighted in Figure 1-63.

A 90° flange is previewed.

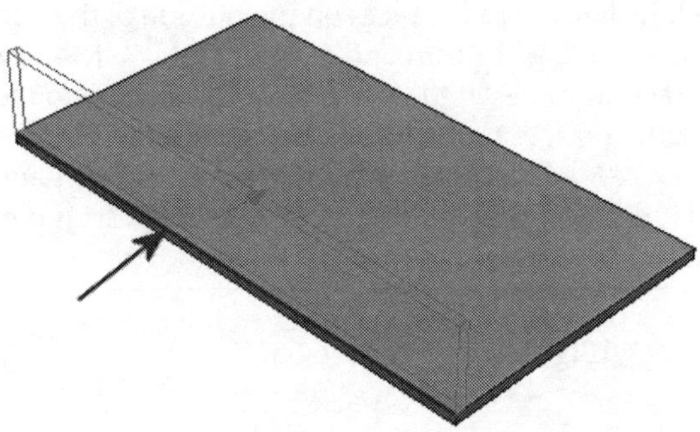

Figure 1-63

6 In the Flange dialog box:

   a. Enter **47.5 mm** in the Distance edit box.

   b. Enter **72 deg** in the Angle edit box.

   c. Enter **12 mm** in the Bend Radius edit box. This matches one of the bend radii in the bend table.

   d. Click the Unfold Options tab.

   e. Uncheck Use Default settings.

   f. Select Bend Table from the Unfold Method list. The only available bend table in the sheet metal style is the With Grain named table.

   g. Click Apply.

The part should match the one shown in Figure 1-64.

Figure 1-64

NOTE You can test the accuracy of the bend table by sketching on the end face of the part. Extend the lines from the faces on the outside of the bend. The measured lengths A and B are 84.012 mm and 50.696 (Total = 134.708). An interpolation of the correction values in the bend table (12 mm bend radius column) returns a bend length correction of 4.92 mm. The unfolded length is 134.708–4.92 = 129.788 mm. Create a flat pattern of the part and place a 1:1 scale view of the flat pattern. The length of the unfolded part is dimensioned as 129.789 mm.

---

7 Click the edge highlighted in Figure 1-65.

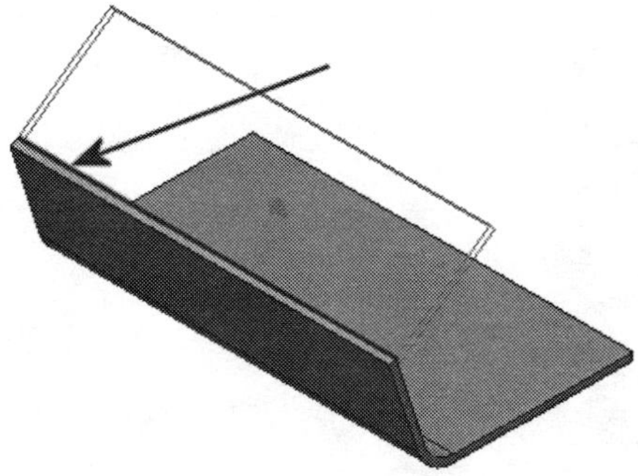

Figure 1-65

8 In the Flange dialog box:

a. Enter **67 mm** in the Distance edit box.

b. Enter **108 deg** in the Angle edit box.

c. Click the Flip Direction tool located to the right of the Distance edit box.

The previewed flange should match the one shown in Figure 1-66.

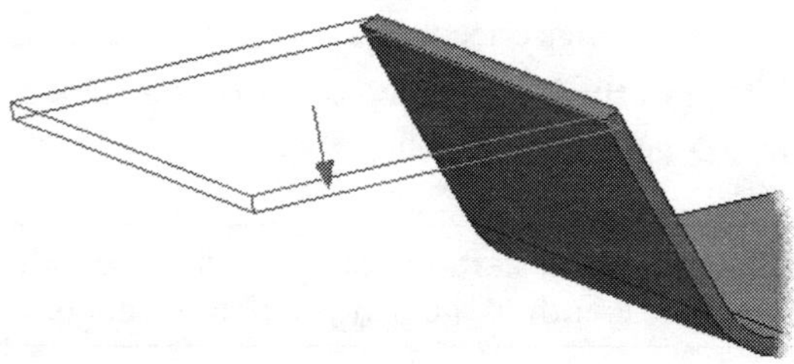

Figure 1-66

9 Click Apply.

Creating Bends using the KFactor

You now create a flange with a bend across the sheet, using the default sheet metal style KFactor.

1 Click on the edge highlighted in Figure 1-67.

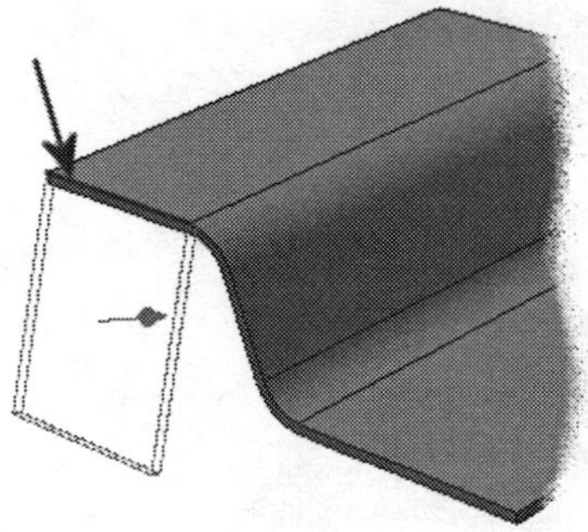

Figure 1-67

2 In the Flange dialog box:
   a. Enter **34 mm** in the Distance edit box.
   b. Enter **90 deg** in the Angle edit box.
   c. Click the Flip Direction tool located to the right of the Distance edit box.
   d. Select BendRadius from the Bend Radius list.
3 Click the Unfold Options tab.
4 Check the Use Default Settings checkbox.
5 Click the More button in the lower right corner of the Flange dialog box.
6 Select Width from the Type list.

NOTE The Width option requires a starting point, an offset from the point to the start of the flange, and the length of the flange along the edge.

---

7 Click on the endpoint highlighted in Figure 1-68.

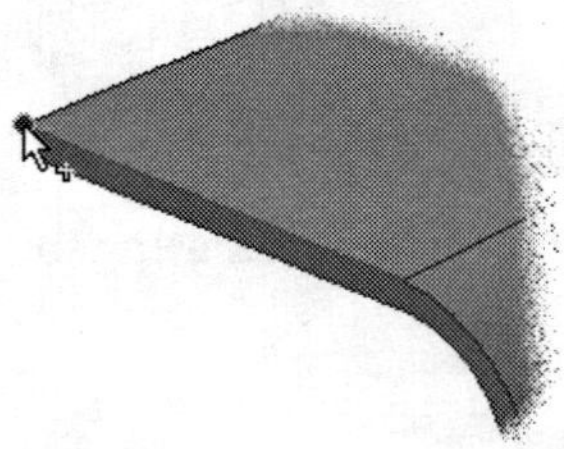

Figure 1-68

8 In the Extents area of the Flange dialog box:

   a. Enter **5 mm** in the Offset edit box.

   b. Enter **40 mm** in the Width edit box.

   c. If the flange previews in the wrong direction, click the Flip Direction tool between the Offset and Width edit boxes.

The flange preview should match the one shown in Figure 1-69.

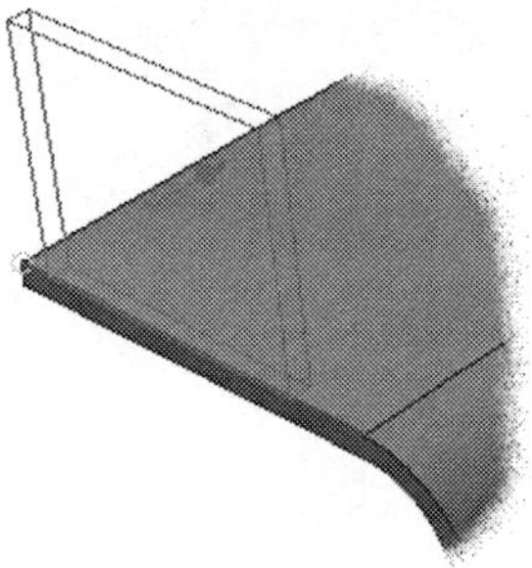

Figure 1-69

9 Click Apply.

Add an Additional Flange

The Flange tool should still be active.

1 Click the edge highlighted in Figure 1-70.

The Width option for flange extents is still active.

2 Click the endpoint highlighted in Figure 1-70.

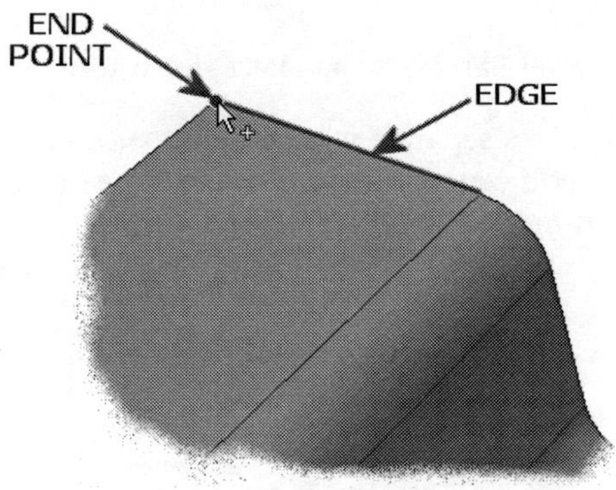

Figure 1-70

3 Click the Flip Direction tool in the Extents area.

4 In the Bend area of the Flange dialog box, uncheck the Use Bend Relief checkbox.

5 Click OK.

Your part should match the one shown in Figure 1-71. The flange tears the adjacent face in the bend zone.

Figure 1-71

6 Select the face highlighted in Figure 1-72.

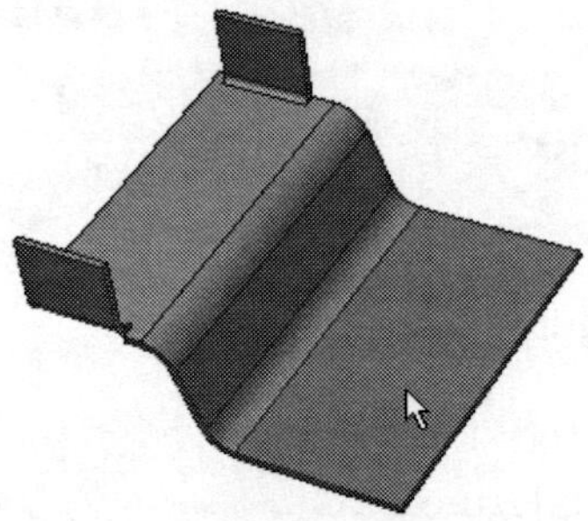

Figure 1-72

7 From the Sheet Metal Panel Bar, click the Flat Pattern tool.

The flattened sheet is shown in a separate window, and should match the one shown in Figure 1-73. The bend centerlines and the extents of the bend zones are indicated on the flat pattern model.

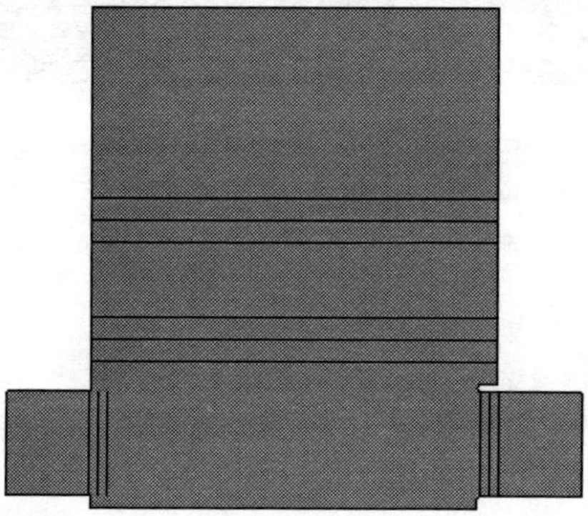

Figure 1-73

8 Close the assembly without saving.

End of exercise.

Exercise 3: Rolled Sheet Metal Parts

Autodesk Inventor can unfold some rolled sheet metal parts such as cylinders and concentric cones. The following short exercise demonstrates the creation of a flat pattern for a concentric cone.

1 Open the file *SM-Cone.ipt*.

The part consists of a single sketch as shown in Figure 1-74.

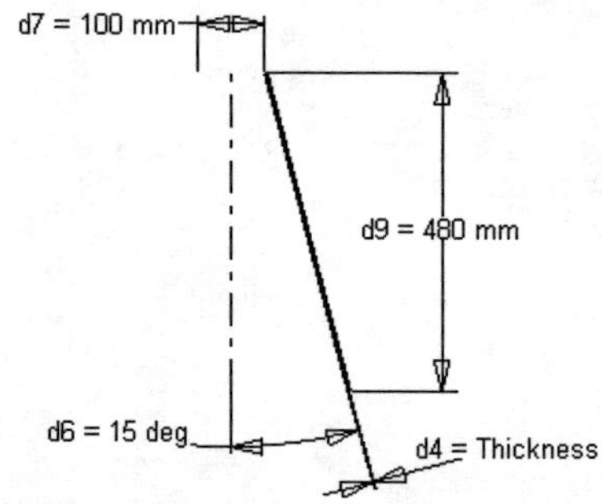

Figure 1-74

2 Right-click in the Sheet Metal Panel Bar, and select Features from the menu.

3 From the Features Panel Bar, click the Revolve tool.

NOTE Since the sketch consists of a single closed profile and a centerline, the profile and axis selections are preselected.

---

4 In the Revolve dialog box:

a. Select Angle from the Extents list.

b. Enter **359.9 deg** in the Angle edit box.

c. Click OK.

5 Right-click in the Features Panel Bar, and select Sheet Metal from the menu.

6 In the Sheet Metal Panel Bar, click the Flat Pattern tool.

NOTE A message box is displayed, indicating that the flat pattern could not be generated.

---

7 Click Cancel.

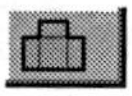

8 Select the outer face of the cone, and then click the Flat Pattern tool in the Sheet Metal Panel Bar. The flat pattern model is successfully generated, as shown in Figure 1-75. In addition, a Flat Pattern node is added to the Browser.

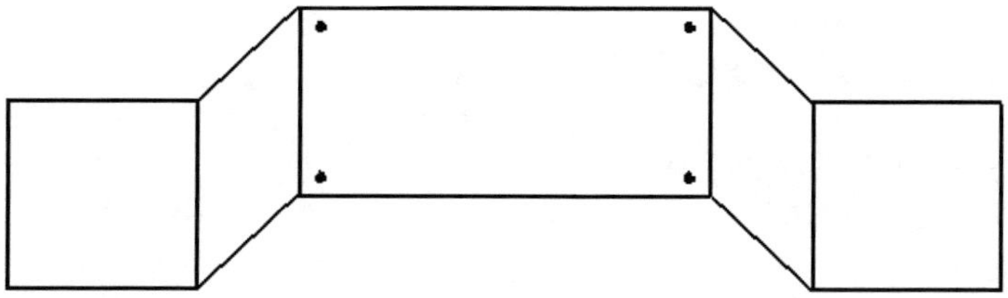

Figure 1-75

Working with Flat Patterns

The flat pattern has no edge aligned horizontally or vertically.

1 Right-click FlatPattern in the Browser.
2 Select Align Horizontal from the menu.
3 Select one edge of the split as shown in Figure 1-76.

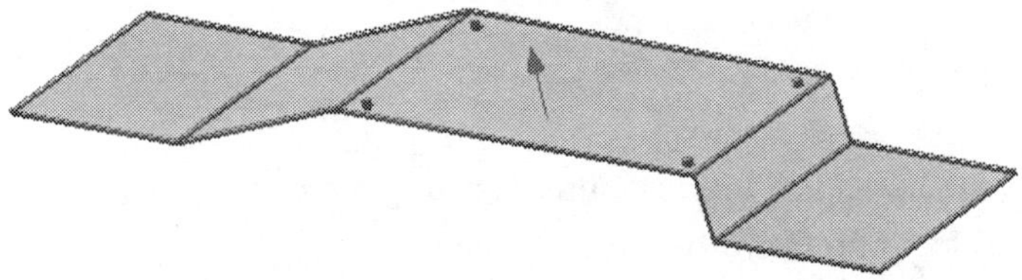

Figure 1-76

From the context menu, you can obtain the flat pattern extents (XY bounding box), select a new fixed face in the folded model, and export the flat pattern in DWG, DXF, and SAT formats.

TIP You cannot use the measure tool in the flat pattern window. The Extents option from the context menu displays the dimensions of the smallest rectangle (aligned to the *XY* axes) enclosing the flat pattern. To measure edge lengths, perimeter, and area of the flat pattern, export the flat pattern as a SAT file. Open the SAT file in Autodesk Inventor and then use the measure tool to obtain these values.

---

4 Close the file without saving changes.

End of exercise.

Exercise 4: Working with Imported Geometry

In this exercise, you will import a flat pattern drawing and create a folded sheet metal part using the Fold tool.

Open an AutoCAD Drawing

1 Click the Open tool.

2 Select AutoCAD Drawing (*.dwg) from the Files of type list as shown in Figure 1-77.

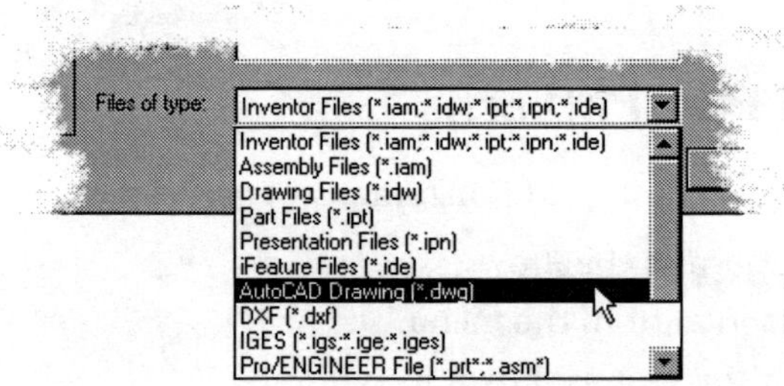

Figure 1-77

3 Select *FlatPatt.dwg*.

4 Click Options.

5 In the DWG File Import Options dialog box:

a. Select mm from the Units of File list.

b. Click Next.

6 In the Layers and Objects Import Options dialog box:

a. Select Read Content From Model Space.

b. Clear the check mark from the Dimension Layer.

c. Check the Constrain endpoints option.

d. Click Next.

7 In the Import Destination Options dialog box:

a. Select the New Part Option.

b. Click the Browse button next to the Part Template edit box.

c. Select the *Sheet Metal (mm).ipt* template from the Metric tab.

d. Click OK.

e. Click Finish.

8 Click Open.

9 Click the Zoom All tool.

A sheet metal part is opened with a single sketch containing the flat pattern layout of the part as shown in Figure 1-78.

The four vertical lines define the centerlines of the required bends.

Each side requires a 30° bend up next to the center section, and a corresponding 30° bend down on the outer bend line.

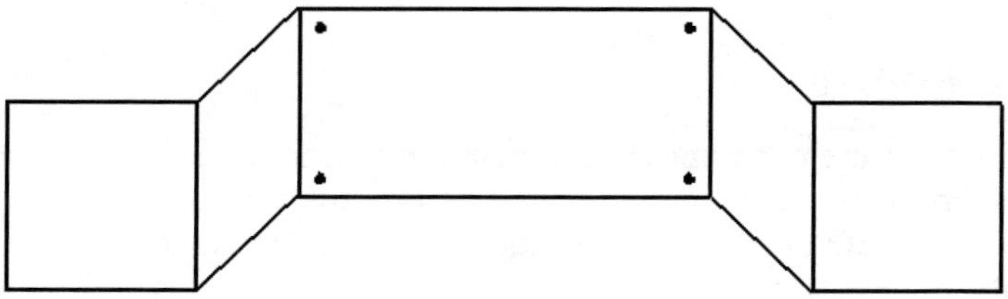

Figure 1-78

The endpoints of all lines are coincident to other endpoints or to lines, but there are additional sketch constraints required to fully constrain the sketch.

If the flat pattern might require changes later, you can add sketch constraints and dimensions to fully constrain the sketch.

Flat Shape

Next, you create a flat blank from the imported sketch geometry and then use the Fold tool to create the formed part.

1 Click the Sheet Metal Styles tool from the Sheet Metal Panel Bar.

2 Enter **2.5 mm** in the Thickness edit box.

3 Select Steel, Mild from the Material list.

4 Click Save.

5 Click Done.

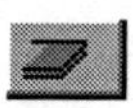

6 Click the Face tool from the Sheet Metal Panel Bar.

7 Click inside the five closed profiles as shown in Figure 1-79.

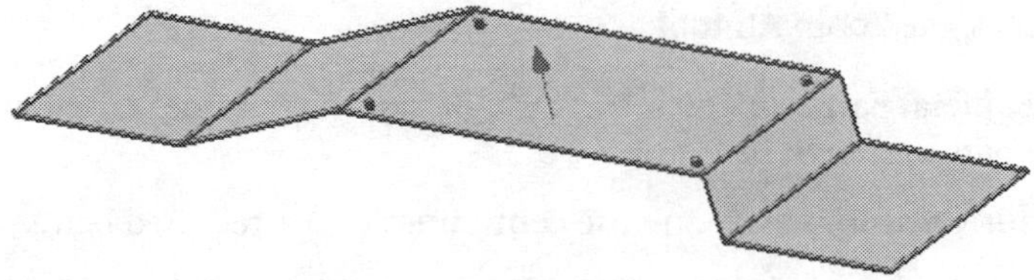

Figure 1-79

8 Click OK.

Fold Features

The Face feature consumes the sketch containing the fold lines. You might think that sharing the sketch would be the best method of obtaining the fold lines. The folds will fail if you use this technique since the first sketch is located on the origin XY plane.

Fold lines must be from a sketch on a part face. You will create a new sketch on the top face of the blank, and project the fold lines from the first sketch.

1 Start a new sketch on the face highlighted in Figure 1-80.

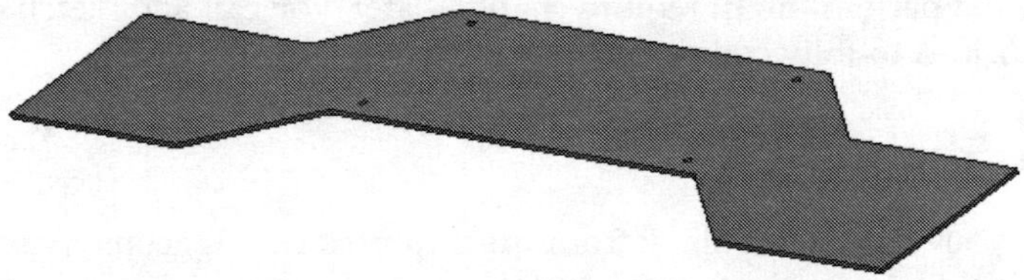

Figure 1-80

2 Expand Face1 in the Browser.

3 Right-click Sketch1 and select Visibility from the menu.

4 Switch to Wireframe display.

5 Click the Project Geometry tool from the Sketch Panel Bar.

6 Project the four edges shown in Figure 1-81.

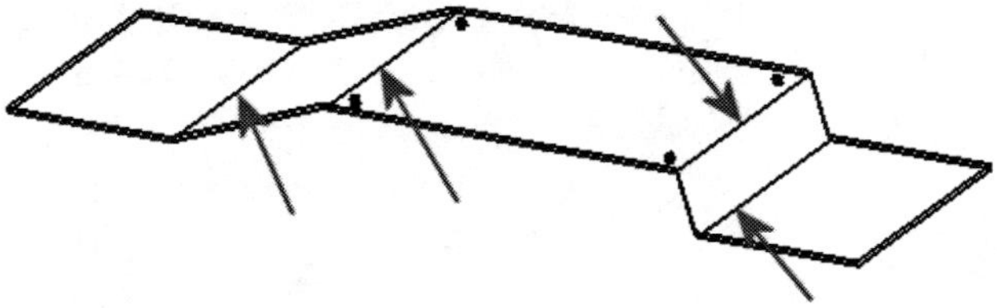

Figure 1-81

7 Finish the sketch.

8 Change to a Shaded display.

9 Turn off the visibility of Sketch1.

Add Fold Features

1 Click the Fold tool from the Sheet Metal Panel Bar.

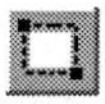

2 Click the Zoom Window tool.

3 Zoom in on the two bend lines on the right side of the blank.

4 Click the right most bend line as shown in Figure 1-82.

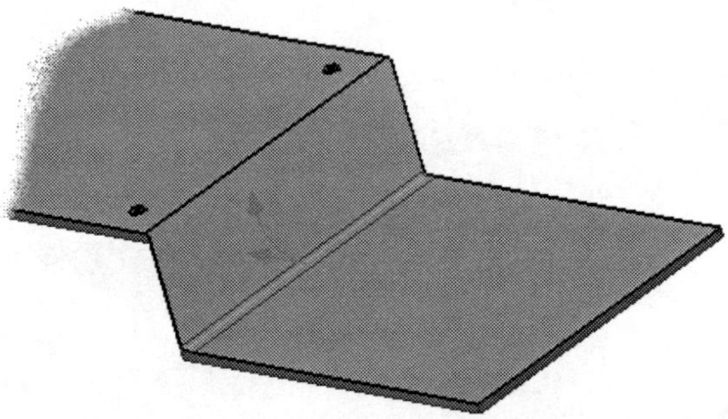

Figure 1-82

The arrows preview the current settings for the bend. The face opposite the arrows will remain fixed, and the face on the other side of the bend line will bend up. The bending zone is the area between the red lines.

5 In the Fold dialog box:

a. Click the Flip Side tool.

b. Click the Flip Direction tool.

The fold preview should match the one shown in Figure 1-83.

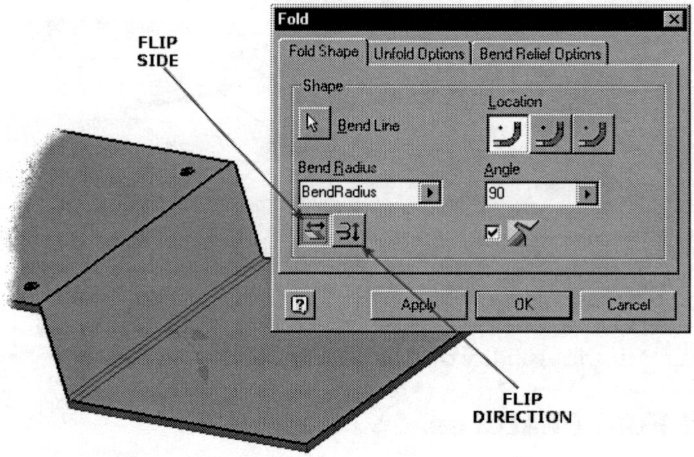

Figure 1-83

6 Enter **30** in the Angle edit box.
7 Click OK.

The blank is bent down at the fold line as shown in Figure 1-84.

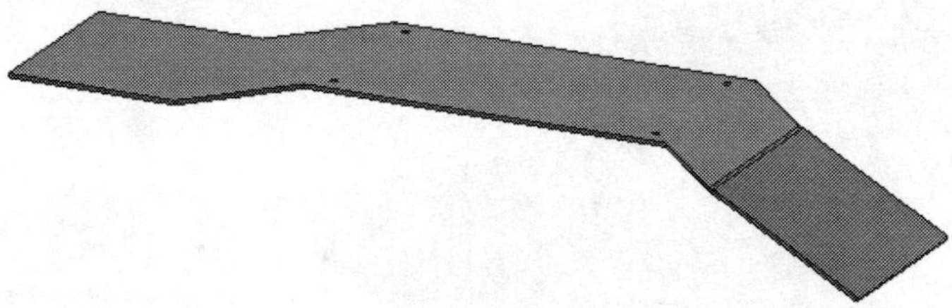

Figure 1-84

8 Expand Fold1 in the Browser.
9 Right-click Sketch2 and select Share Sketch from the menu.

10 Click the Fold tool from the Sheet Metal Panel Bar.
11 Click the bend line shown in Figure 1-85.

NOTE If the preview arrows do not match those shown in Figure 1-85, use the Flip Side and Flip Direction tools to change the fold orientation.

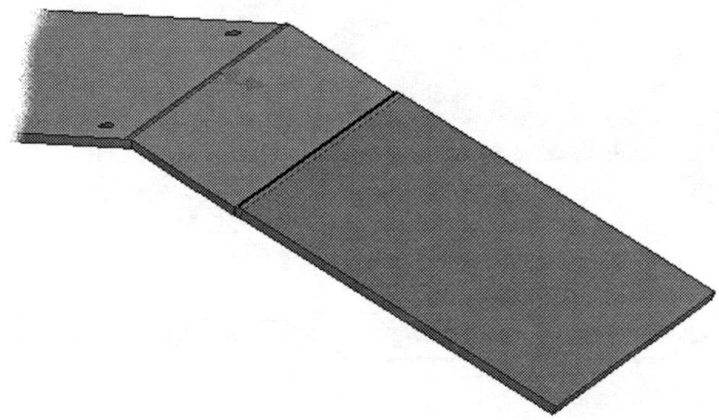

Figure 1-85

12 Set the fold angle to 30°.

13 Click OK.

Your part should match the one shown in Figure 1-86.

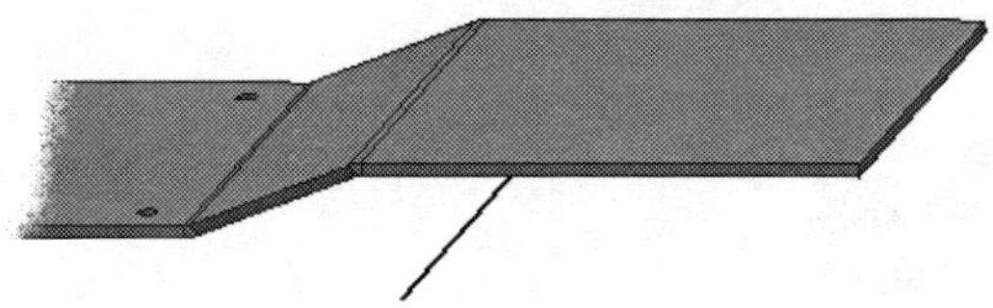

Figure 1-86

14 Use the Fold tool to create the same bends on the opposite end of the blank.

Your part should match the one shown in Figure 1-87 when you have completed the four folds.

15 Turn off the visibility of Sketch2.

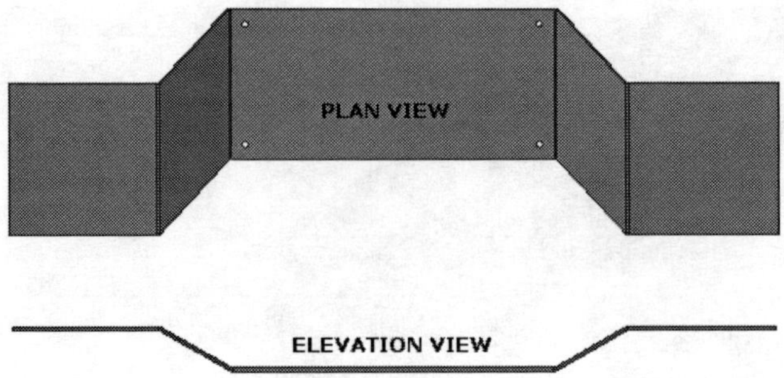

Figure 1-87

Add Additional Folds

Next, you add a fold on one of the outer flanges. This fold is at an angle to the flange edges.

1 Start a new sketch on the face highlighted in Figure 1-88.

Figure 1-88

2 Sketch and dimension the line shown in Figure 1-89.

NOTE The line endpoints must be coincident to the face edges.

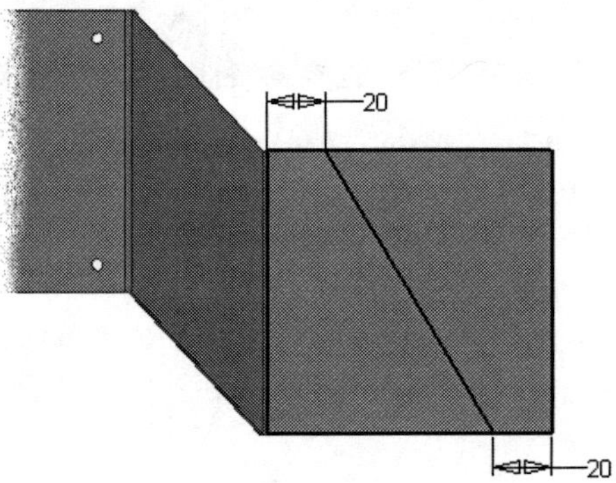

Figure 1-89

3 Finish the sketch.

4 Select the Fold tool from the Sheet Metal Panel Bar.

5 Change the bend line location to be the start of the bend.

6 Create a 90° fold to match the one shown in Figure 1-90.

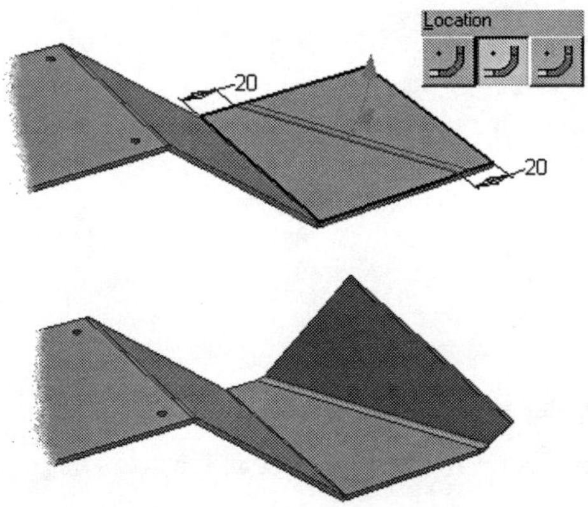

Figure 1-90

7 Save the file as *Fold.ipt*.

You will use this part in a later exercise.

End of exercise.

Exercise 5: Importing a Folded Model

It is common to receive part models from outside sources that are not Autodesk Inventor native files. IGES and STEP formats are common neutral file formats. If the part is imported in a folded state, you can create a flat pattern if all faces are the same thickness

In this exercise, you will import a folded part as a step file. You then edit the sheet metal style to match the part geometry, and then unfold the part.

1 Click the Open tool.
2 Select STEP (*.stp, *.ste, *.step) from the Files of type list.
3 Select *FoldStep.stp* and click Open.

The part should match the one shown in Figure 1-91.

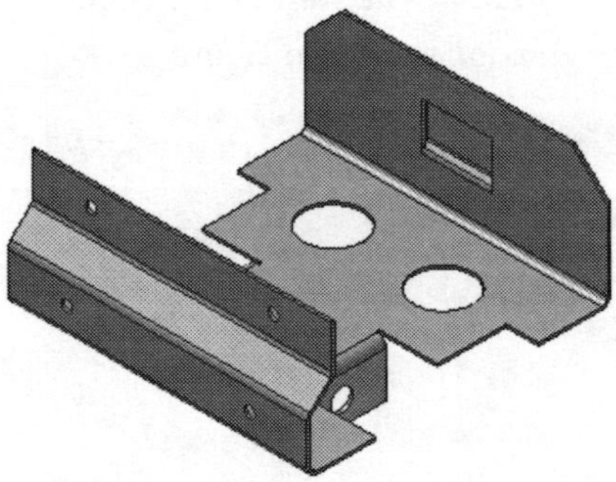

Figure 1-91

4 Select Application > Sheet Metal from the main menu.
5 Right-click in the graphics window and select Measure Distance from the menu.
6 Measure the thickness of the part at a corner as shown in Figure 1-92.

NOTE If your Standard part template has inches as the linear document units, the sheet thickness will be reported as 0.051 in.

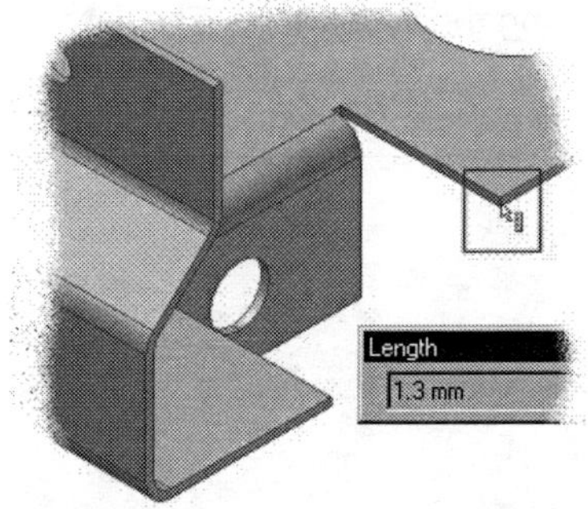

Figure 1-92

7 Click the Styles tool from the Sheet Metal Panel Bar.

8 Enter **1.3 mm** in the Thickness edit box.

TIP You can measure the inside radius of a bend to obtain the typical bend radius for the part. Change the BendRadius value in the Sheet Metal Style to match this measurement. Any additional bends will then match the existing bends.

---

9 Click Save.

10 Click Done.

11 Click the Flat Pattern tool from the Sheet Metal Panel Bar.

A flat pattern of the part is shown in a separate window (see Figure 1-93).

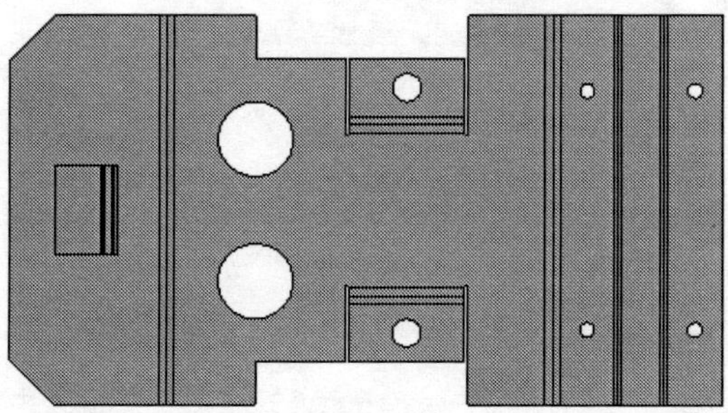

Figure 1-93

NOTE The default Unfold Method Value (KFactor or Bend Table) is used to determine the unfolded face lengths. For an accurate flat pattern, ensure you have the correct value set as the default in the current Sheet Metal Style.

TIP Preselect a key face before generating the flat pattern. The flat pattern analyzer considers a preselected face as fixed, and works outward from this face to unfold the part. Try this technique if a flat pattern fails without a preselected face.

12 Close the file without saving changes.

End of exercise.

iFeatures and Punches

PEM Nut Relief

PEM nuts are fasteners that are pressed into a sheet metal part, usually before folding. The nuts are designed to fit in a prepunched hole. The design of the nut causes the material around the hole to deform and capture the nut in the face. Serrated edges (or other noncircular shape) prevent the nut from spinning in the material. Figure 1-94 shows a PEM F series flush nut placed in a prepunched hole (left side), and the nut after it has been pressed into the face (right side). The hex shape of the nut is larger than the locating hole. As it is pressed into the material, the hex shape displaces material into the groove cut in the nut. This captures the nut in the face, and the hex shape prevents it from turning.

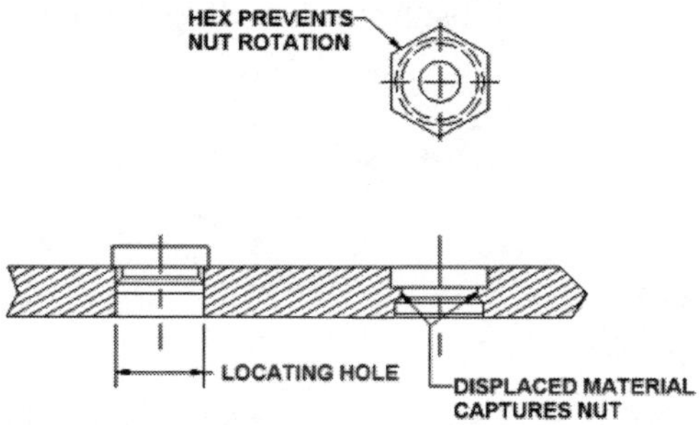

Figure 1-94

The next exercise includes the creation of a punch tool that removes material from the sheet metal face to provide clearance for the hex shape on this style of PEM nut. This is one technique for inserting pressed fasteners in sheet metal parts, but you can use other methods to meet your design requirements.

Method A

PEM fasteners include nuts, standoffs, clips, and other fasteners. Most of these are parts that extend beyond the face they are pressed into (as opposed to the flush nut shown above). You can model the part as an iFeature and place the iFeature on the face of a sheet metal part. The sheet metal component then includes all PEM fasteners in a single part definition. The disadvantages of this technique are that iFeatures cannot include thread information, and the fasteners are not included as components in a BOM or Parts List.

Method B

You create an iFeature or punch to describe the location hole and clearance for the profile pressed into the part face (this technique is used in the following exercise).

You then place the PEM fastener in the assembly. This technique captures the fasteners in the BOM or Parts List, and provides for an assembly with no interference between the fastener and the sheet metal part. The disadvantage is that the volume of the sheet metal part is incorrect because the nut displaces (does not remove) material as it is pressed in.

Method C

With this technique, you create the location hole for the fastener, but you do not remove material to provide clearance for the hex head or serrated edges. This technique generates an accurate part volume and flat pattern, but the assembly will report interference between the fastener and the sheet metal part.

Method D

Create modified PEM components that can be used with the simple clearance hole in the sheet metal face. A very thin feature that sits on top of the sheet metal face represents the hex shape or serrated teeth that are pressed into the sheet metal face.

Exercise 6: iFeatures and Punches

In this exercise, you create two iFeatures: a PEM nut relief and a bridge lance. Both iFeatures include geometry to define them as punches. You then use the Autodesk Inventor PunchTool to place instances of these iFeatures in a sheet metal part.

PEM Nut Hex Clearance iFeature

In this section, you create the location hole and hex clearance for a flush mount PEM nut. The hex feature is saved as a punch.

1 Open *iFeatureBase.ipt*.

The part should match the one shown in Figure 1-95.

Figure 1-95

This part is a simple sheet metal face that will serve as the base for iFeature creation. Being a sheet metal part, it contains the sheet metal parameters such as Thickness and BendRadius required for proper sheet metal iFeature definition.

2 Create a sketch on the top surface of the face.

3 Sketch and dimension a 4.35 mm diameter circle near the position of the hole shown in Figure 1-97.

4 Finish the sketch and use the Cut tool from the Sheet Metal Panel Bar to cut the circle through the sheet metal face.

The part should match the one shown in Figure 1-96.

Figure 1-96

Add Additional Sketch Features

1 Create a new sketch on the same face.

2 Use window select or crossing window select to select all projected geometry on the face. Delete all projected geometry.

TIP It is good practice to delete all automatically projected geometry in the first sketch of an iFeature. All geometry in the first sketch is displayed when placing an iFeature or punch.

3 Click the Project Geometry tool from the Sketch Panel Bar.

4 Project the edge of the 4.35 mm diameter circle into the sketch.

5 Click the Point, Hole Center tool.

6 Place a Point, Hole Center at the center of the hole.

NOTE The Hole Center enables the iFeature to be placed as a punch feature.

7 Click the Polygon tool on the Sketch Panel Bar.

8 Create a 6-sided polygon centered on the hole.

9 Click the Dimension tool on the Sketch Panel Bar.

10 Dimension the polygon across the flats and enter **4.82 mm** in the dimension edit box.

Your sketch should match the one shown in Figure 1-97.

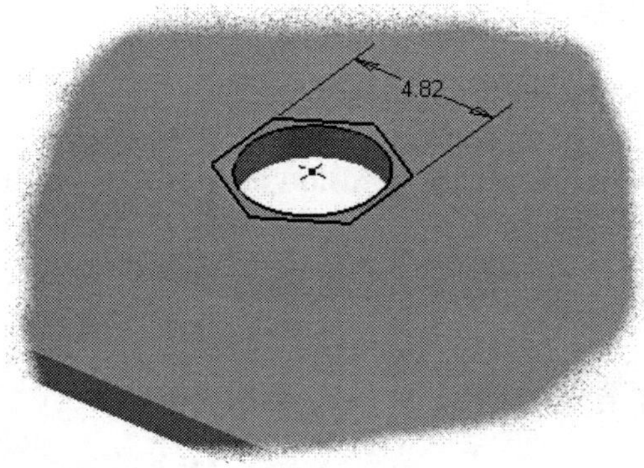

Figure 1-97

11 Click the Extrude tool from the Features Panel Bar.

TIP You can also press the E key to start the Extrude tool.

---

12 Click between the polygon and the projected circle to select the extrude profile as shown in Figure 1-98.

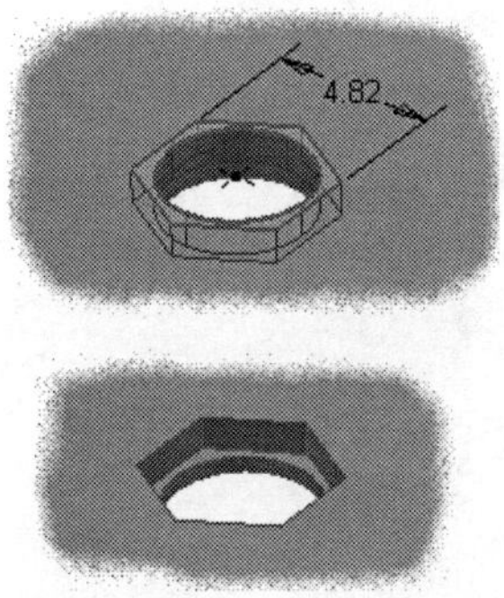

Figure 1-98

13 In the Extrude dialog box:

a. Select the Cut tool.

b. Enter **1 mm** as the extrude distance.

c. Ensure the cut extrudes into the face.

d. Click OK.

Extract the iFeature

1 Click the Parameters tool on the Standard toolbar.

2 In the Parameters dialog box edit the parameter names associated with the dimension across the hex and hex extrusion depth to match those shown in Figure 1-99.

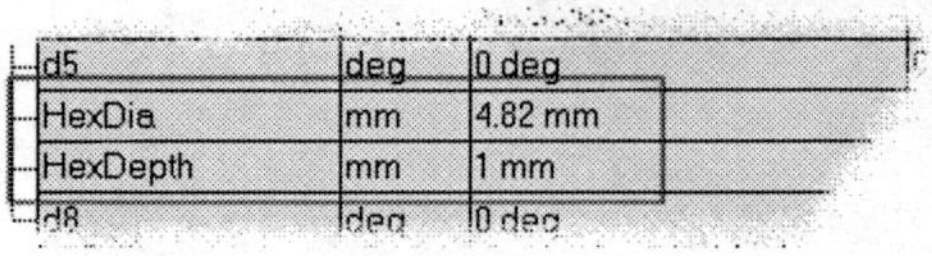

Figure 1-99

3 Click Done.

4 From the Sheet Metal Panel Bar, click the arrow next to the View Catalog tool and then click the Create iFeature tool.

5 Click the Extrusion feature in the Browser.

The required parameters for the iFeature are shown in the Create iFeature dialog box.

6 Drag the HexDia size parameter above the HexDepth size parameter.

7 Edit the size parameter prompts to match those shown in Figure 1-100.

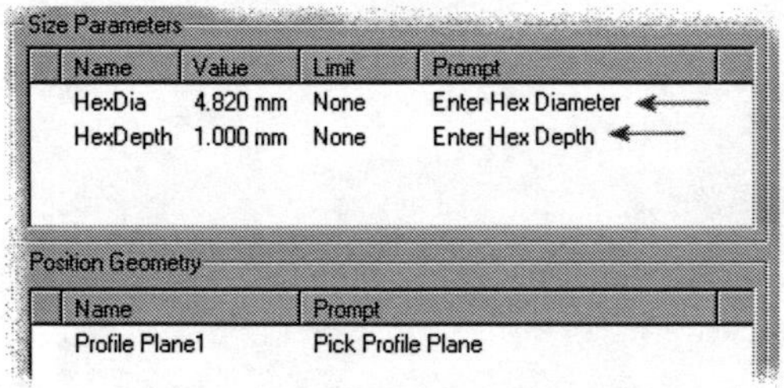

Figure 1-100

8 Click Save.

9 Browse to the Punches folder and enter *PEM-F.ide* as the file name.

10 Click Save.

Bridge Lance Punch

A bridge lance is a common sheet metal punch feature similar to the one shown in Figure 1-101. The punch tears the sheet metal face and deforms the torn area into a die to create a bridge on the underside of the face. These features are often used to contain wiring or provide mounting points for clips.

In the following section, you examine an existing bridge lance feature, and then extract it as a punch iFeature.

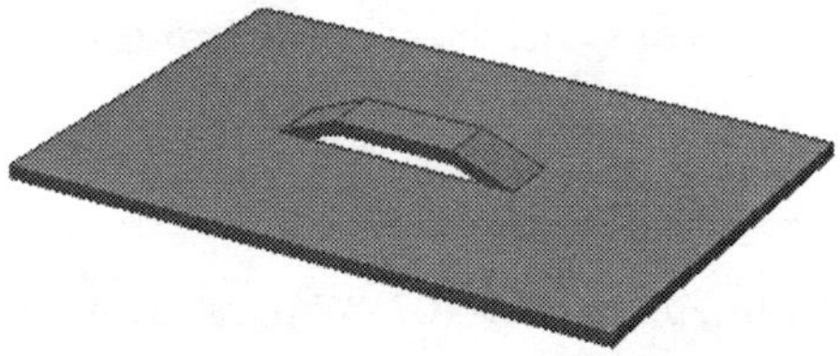

Figure 1-101

1 Open *BridgeLance.ipt.*
2 Right-click Cut1 in the Browser.
3 Select Edit Sketch from the menu.

4 Click the Show Constraints tool from the Sketch Panel Bar.
5 Display the geometric constraints for the four lines of the rectangle.

Your view should match Figure 1-102.

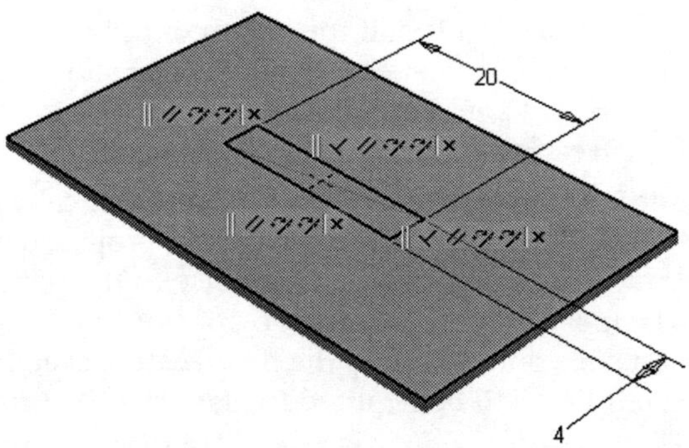

Figure 1-102

NOTE There are no Horizontal or Vertical constraints on the sketch. The rectangle is aligned with the face edges, but the Horizontal constraint was removed.

The saved iFeature will be more reliable when placed at an angle if you remove Horizontal and Vertical constraints in the first sketch contained in the iFeature.

In addition, the Hole Center entity located at the center of the rectangle enables the extracted iFeature to be placed using the PunchTool.

---

6 Exit the sketch.

7 Edit the sketch under Extrusion1.

8 Right-click in the graphics window and select Slice Graphics from the menu.

9 Right-click one of the dimensions and select Show Expression from the menu.

NOTE You may already be viewing dimensions in this method.

Your view should match the one in Figure 1-103 (Wireframe Display).

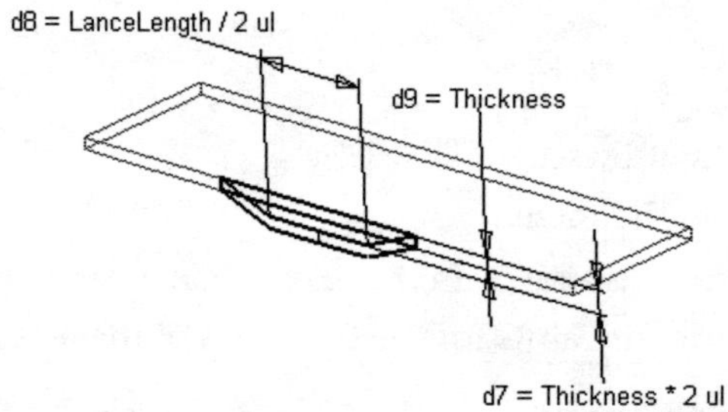

Figure 1-103

a. The length of the flat section is half the length of the cut. The depth of the lance is related to the face thickness. You can create any relationship between parameters, but any stand-alone value must be included in the size parameters of the iFeature.

b. The sketch plane is a face created by Cut1. All automatically projected edges were first deleted, and then only the top edge was projected into the sketch. This is important because it removes a reference to the sketch face that would need to be included in the iFeature definition. Since the top edge can be considered an edge on the face of the first iFeature sketch (Cut1 sketch), a single placement face will be required to place the iFeature or punch.

c. The extrusion distance of the feature is equated to the width of the Cut1 rectangle.

10 Finish the sketch.

11 Click the Parameters tool on the Standard toolbar.

12 Examine the relationship between parameters.

13 Click Done.

Extract the iFeature

1 From the Sheet Metal Panel Bar, click the arrow next to the View Catalog tool and then click the Create iFeature tool.

2 Click Cut1 in the Browser.

The required parameters for the iFeature are shown in the Create iFeature dialog box.

NOTE A single plane selection is required to place the iFeature.

3 Edit the Size Parameter prompts to match those shown in Figure 1-104.

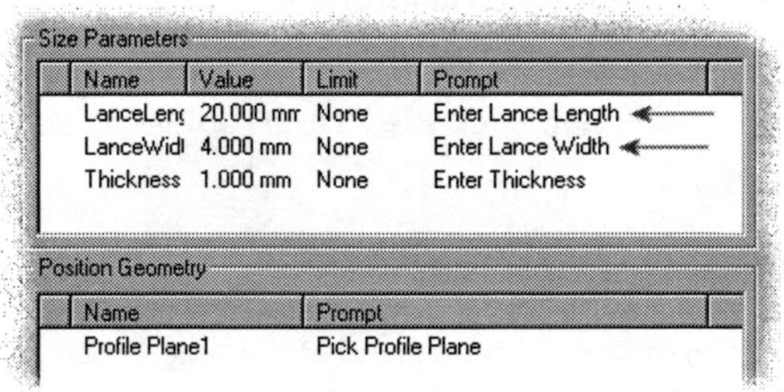

Figure 1-104

4 Click Save.

5 Browse to the Punches folder and enter *BridgeLance.ide* as the filename.

6 Click Save.

Placing Punch iFeatures

Next, you use the PunchTool tool to place instances of the two iFeatures created in the previous exercise.

1 Open *Fold.ipt*.

2 Start a new sketch on the face highlighted in Figure 1-105.

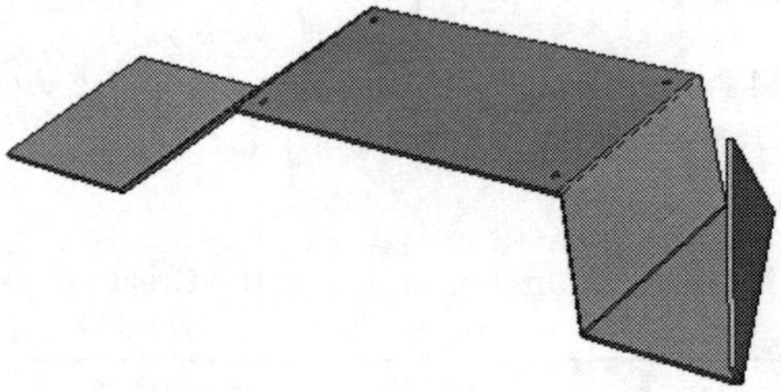

Figure 1-105

3 From the Sketch Panel Bar, click the Point, Hole Center tool.

4 Place hole centers on the center points of the four holes of the face.

5 Finish the sketch.

6 From the Sheet Metal Panel Bar, click the PunchTool tool.

7 Select *PEM-F.ide*.

8 Click Next.

NOTE Four instances of the punch are previewed at the hole center marks. No other position inputs are required to orient the punch features.

---

9 Click Next.

The size parameters are displayed in the final window.

10 Click OK to place the punch features.

Your model should match the one shown in Figure 1-106.

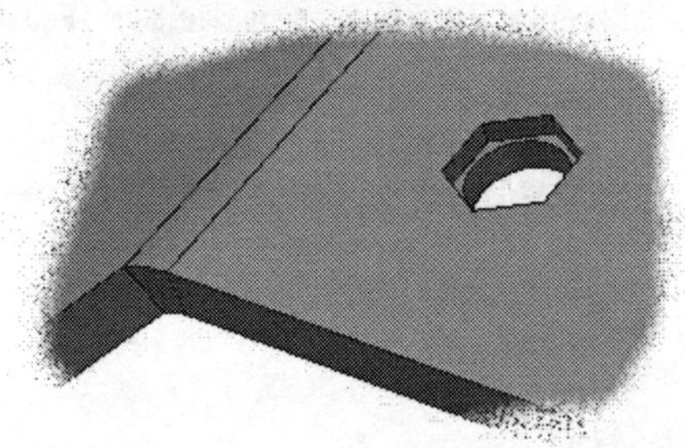

Figure 1-106

Place the Sheet Metal Punch

1 Start a new sketch on the face highlighted in Figure 1-107.

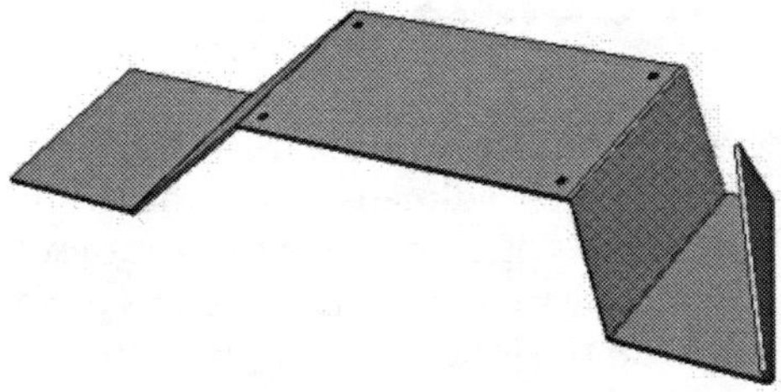

Figure 1-107

2 Click the Point, Hole Center tool.
3 Place a Point, Hole Center near the middle of the face.
4 Finish the sketch.

5 From the Sheet Metal Panel Bar, click the PunchTool tool.
6 Select the BridgeLance.ide punch.
7 Click Next in the PunchTool dialog box.
8 Enter **45 in.** in the Angle edit box, and then click Next.
9 Enter **25 mm** as the LanceLength value.
10 Enter **7.5 mm** as the LanceWidth value.
11 Click OK to place the punch feature.

Your part should match the one shown in Figure 1-108.

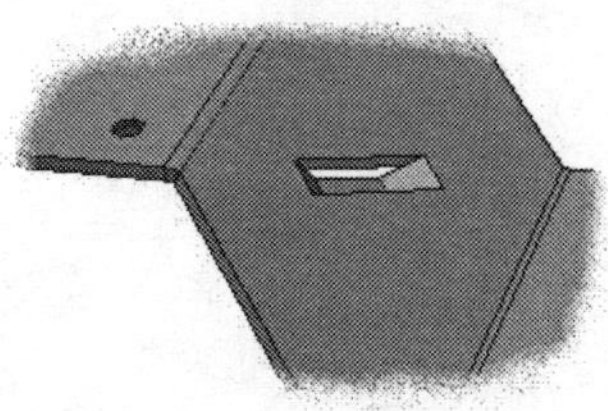

Figure 1-108

12 Close the files without saving changes.

End of exercise.

Sheet Metal Parts from Solids

Some folded sheet metal parts are difficult to create using the face and flange technique. If you do not know the flat pattern dimensions, using the fold tool to create the bend up part is not an option. It is often possible to create a solid model of the part using the standard part modeling tools, and then use this part to create a sheet metal part that can be unfolded. In the following exercises, you examine two techniques for creating sheet metal parts from solid objects:

- Shell the solid object, and then rip open key edges to enable the part to unfold. You then add bends at the remaining sharp edges.
- You derive one or more surface bodies into the sheet metal part, these surfaces represent the inside or outside surface of a sheet metal face. The sheet metal part is created from the derived surfaces.

Exercise 7: Shelled Solids

You can build sheet metal part from shelled solid parts. The part can be a native Autodesk Inventor part model, or an imported base solid.

You rip open appropriate edges to create an unfoldable part, and add bends to other sharp edges. This technique works best with solids consisting of block or other simple shapes.

NOTE This technique should be used when the other modeling techniques cannot.

---

1 Open *Cross.ipt*. The part should resemble Figure 1-109.

Figure 1-109

The part is a Base solid imported from a .sat file. You are to create a sheet metal part with minimal gaps from 1.5 mm sheet stock.

2 Select Application > Sheet Metal from the main menu.

- **a.** Click in the Sheet Metal Panel Bar and select Features from the menu.
- **b.** From the Features Panel Bar, click the Shell tool.
- **c.** Click the top face of the part to remove it from the shell.
- **d.** Click the arrow to the right of the Thickness edit box, and select List Parameters from the menu.
- **e.** Select Thickness and then click OK in the Shell dialog box.

The part is shelled as shown in Figure 1-110.

Figure 1-110

Next, you adjust the Sheet Metal Style to meet the design requirements.

3 Click the Features Panel Bar and select Sheet Metal from the menu.

4 From the Sheet Metal Panel Bar, click the Styles tool.

5 Enter **1.5 mm** in the Thickness edit box.

6 Click the Corner tab. Select Tear from the Relief Shape list as shown in Figure 1-111.

7 Click Save and then click Done.

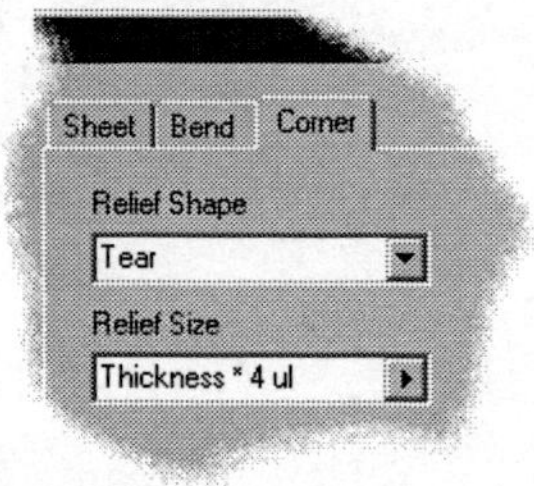

Figure 1-111

Rip the Corners

The best workflow for creating an unfoldable part from a shelled solid is as follows:

1 Rip all required corners.

2 Add bends where appropriate.

3 Add corner seams to clean up corner intersections.

4 Reorient your view to match Figure 1-112.

Figure 1-112

5 From the Sheet Metal Panel Bar, click the Corner Seam tool.
6 In the Corner Seam dialog box, check the Corner Rip option.
7 Click the edge highlighted in Figure 1-113.

Figure 1-113

8 Click Apply.
9 Apply the same corner rip to the three similar edges at the intersection of the arms as shown in Figure 1-114.

TIP You can select two noncontinuous edges per corner rip.

Figure 1-114

10 Next, rip the two connected edges shown in Figure 1-115.

11 Select both edges, and then click Apply.

Figure 1-115

12 Apply additional corner rips to the similar edges on the other arms.

13 Click Cancel in the Corner Seam dialog box.

Your part should match the one shown in Figure 1-116.

Figure 1-116

Add Bends

The part is now in an unfoldable state. Next, you add bends to the remaining joined edges to enable the part to unfold.

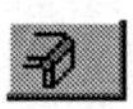

1 From the Sheet Metal Panel Bar, click the Bend tool.

2 Click the horizontal edge highlighted in Figure 1-117, and then click Apply.

3 Repeat for each of the two vertical edges.

Figure 1-117

4 Apply a bend to the same three edges on each of the arms.
5 Add the bend to the horizontal edge first, and then to each of the vertical edges.
6 Click Cancel in the Bend dialog box.

Add Corner Seams

Finally, you add corner seams with a minimal gap to complete the sheet metal part.

1 From the Sheet Metal Panel Bar, click the Corner Seam tool.
2 Select the two edges highlighted in Figure 1-118.

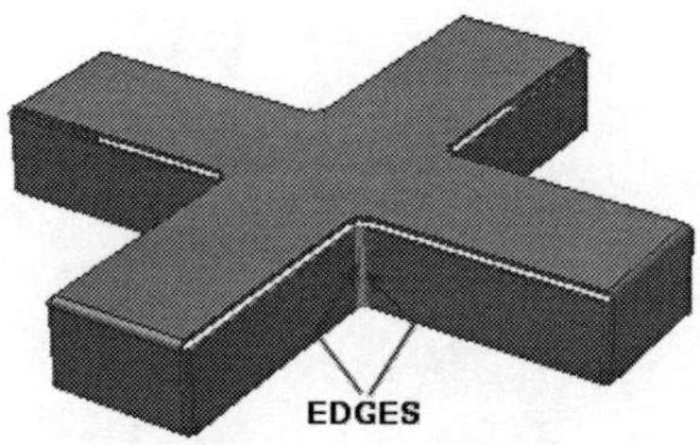

Figure 1-118

3 Enter **.01 mm** in the Gap edit box.
4 Click Apply.
5 Apply the same .01 mm corner seam to the other three edge intersections at the center of the cross.
6 Apply the same .01 mm corner seam to the eight adjacent edge sets represented by the edge selected in Figure 1-119.

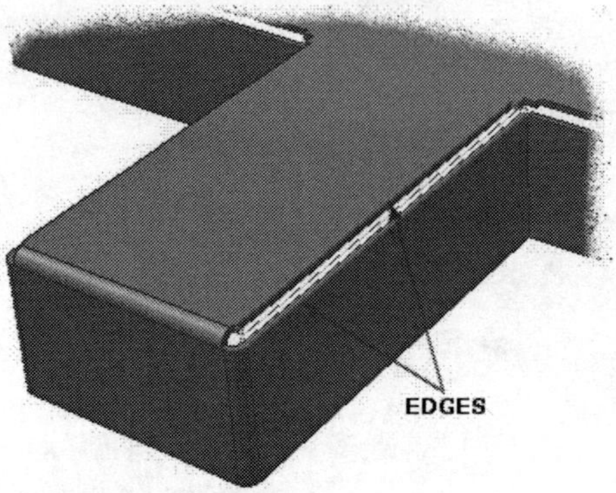

Figure 1-119

7 Click Cancel in the Corner Seam dialog box.

8 Select the bottom face of the cross.

9 Click the Flat Pattern tool from the Sheet Metal Panel Bar.

The flat pattern should match the one shown in Figure 1-120.

Figure 1-120

10 Close the file without saving changes.

End of exercise.

Exercise 8: Lofted Shapes

Transition shapes are difficult to create using the sheet metal tools in Autodesk Inventor. Third party products are available from Autodesk's MAI (Mechanical Applications Initiative) partners that automate the creation of unfoldable sheet metal solutions for square to round and other complex transitions. Within Autodesk Inventor, you can create a linear transition using the loft tool.

By adding appropriate fillets to the loft before the shell operation, the part can be unfolded using the flat pattern tool. However, it is difficult to add flanges and faces to the folded part since the edge faces are not perpendicular to the large sheet metal faces.

You can use the surfaces of the loft by deriving the lofted part into the sheet metal part as a surface body. You can derive any number of surface bodies and sketches into a sheet metal part, and use the geometry to define sheet metal features that remain associative to the underlying geometry.

Complex transitions are not recommended, but this powerful technique extends the range of sheet metal parts that can be created.

1 Open *LoftTrans.ipt.*

The part contains a lofted feature between sketches on two parallel planes. An extrusion extends from the base of the loft.

2 Edit the Loft1 feature.

The smaller rectangular sketch is located near the bottom right corner of the projected larger rectangle as shown in Figure 1-121.

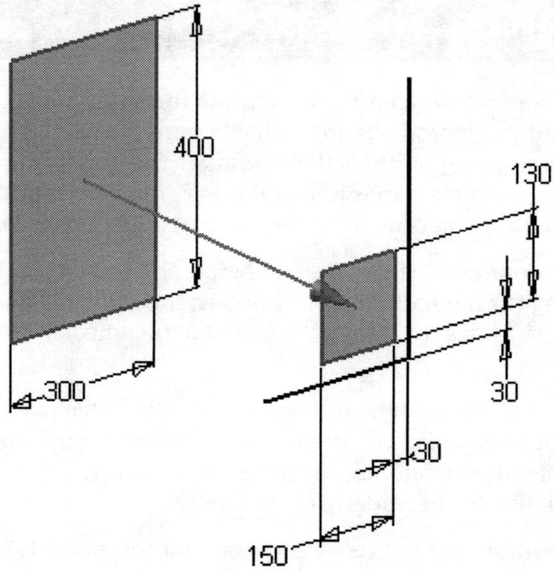

Figure 1-121

3 Finish the edit.

4 Start a new sheet metal part based on the *Sheet Metal (mm).ipt* template on the Metric tab.

5 Exit the initial sketch.

Derive a Component

1 Click in the Panel Bar and select Features from the menu.

2 From the Features Panel Bar, click the Derived Component tool.

3 Select *LoftTrans.ipt* and then click Open.

4 In the Derived Part dialog box, click the Body as Work Surface option as shown in Figure 1-122.

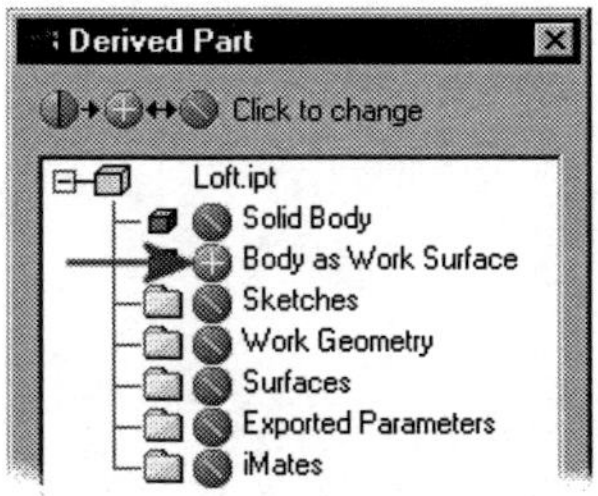

Figure 1-122

5 Click OK.

The outer surface of the lofted shape is imported into the sheet metal part as shown in Figure 1-123. The isometric view of the part is shown.

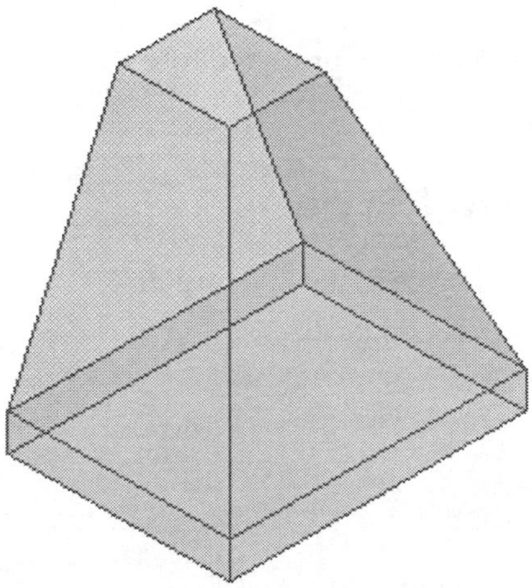

Figure 1-123

6 Change to the current Isometric View.

7 Start a new sketch on the face highlighted in Figure 1-124.

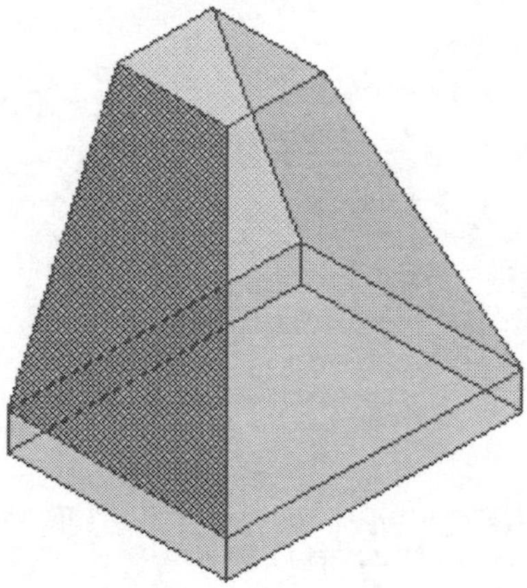

Figure 1-124

The edges of the surface are automatically projected into the sketch.

8 Finish the sketch.

Update the Sheet Metal Style

1 From the Sheet Metal Panel Bar, click the Styles tool.
2 Enter **1.5 mm** in the Thickness edit box.
3 Click Save.
4 Click Done.

Add Sheet Metal Faces

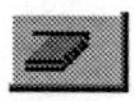

1 From the Sheet Metal Panel Bar, click the Face tool.

Ensure that the sheet metal face extrudes away from the surface body.

2 Click OK.
3 Start a new sketch on the face highlighted in Figure 1-125.

NOTE The edges of the surface are automatically projected into the sketch.

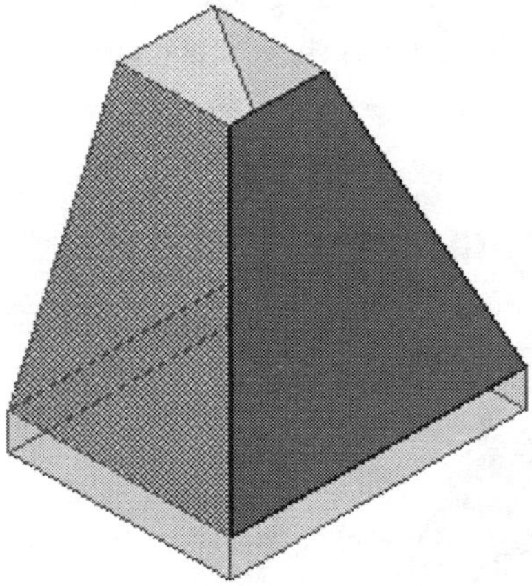

Figure 1-125

4 Click the Line tool from the Sketch Panel Bar.

5 Sketch a line from the midpoint of the upper horizontal edge.

The line is coincident and perpendicular to the bottom edge as shown in Figure 1-126.

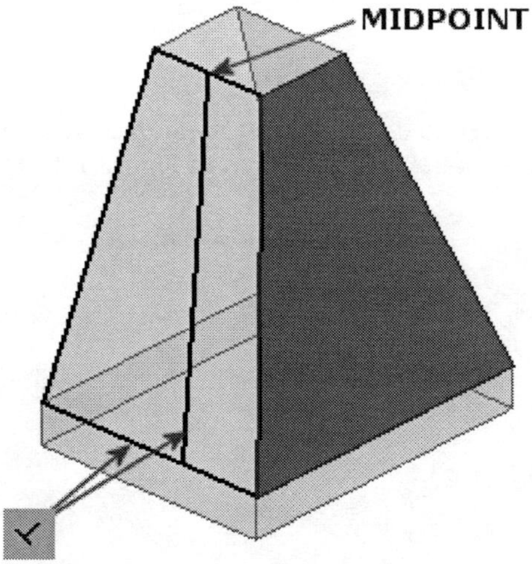

Figure 1-126

6 Exit the sketch.

7 From the Sheet Metal Panel Bar, click the Face tool.

8 Click inside the profile next to the existing sheet metal face. Ensure that the face extrudes away from the surface body.

9 Click the Select Edge tool in the Face dialog box.

10 Click the adjacent edge of the previous sheet metal face as shown in Figure 1-127.

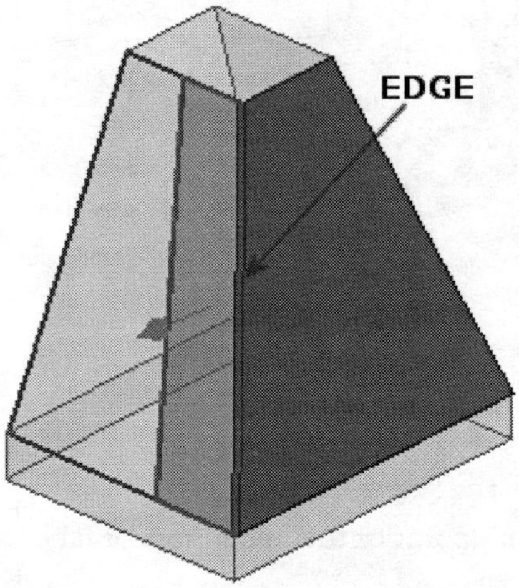

Figure 1-127

11 Click OK.

The new face is joined to the first face with bend.

12 Repeat the above procedure to create a sheet metal face attached to the opposite side of the first sheet metal face as shown in Figure 1-128.

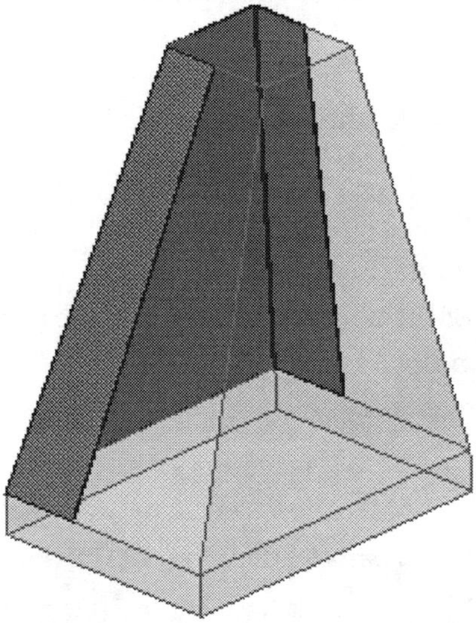

Figure 1-128

13 Save the part as *SM-Trans01.ipt*.

Create a Second Sheet Metal Part

To manufacture the transition, a second sheet metal part is created with an overlap to the first part. You can use both the lofted surface body and the surface body of the first part to help create the second part.

If you create the second component in-place in an assembly, both sheet metal parts can simply be grounded, and their shapes are controlled by the lofted part.

1 Create a new assembly based on the *Standard (mm).iam* template on the Metric tab.

2 From the Assembly Panel Bar, click the Place Component tool.

3 Place a single instance of *SM-Trans01.ipt* in the assembly.

NOTE It is grounded at the assembly origin.

4 From the Assembly Panel Bar, click the Create Component tool.

5 In the Create In-Place Component dialog box:

a. Enter **SM-Trans02** in the New File Name edit box.

b. Click the Browse tool next to the Template edit box.

c. Select the *Sheet Metal (mm).ipt* template from the Metric tab.

d. Clear the check mark from the Constrain sketch plane to selected face or plane checkbox.

e. Click OK.

f. Expand the Origin folder under the top-level assembly in the Browser.

g. Click the XY plane.

A new part is created and activated.

6 Exit from the active sketch in the new part.

7 Exit from the in-place edit of the new part to return to the assembly.

8 Right-click *SM-Trans02.ipt* in the Browser.

9 Select Grounded from the menu.

Derive a Parameter

You can link the thickness of the two parts by deriving the Thickness parameter of the first part into the second part.

You will derive surface bodies from the first sheet metal part into the second part, so we'll set up the parameter to be derived at the same time.

1 Open or edit in-place *SM-Trans01.ipt*.

2 From the Standard toolbar, click the Parameters tool.

3 In the Parameters dialog box check the export parameter checkbox next to the Thickness model parameter, as shown in Figure 1-129.

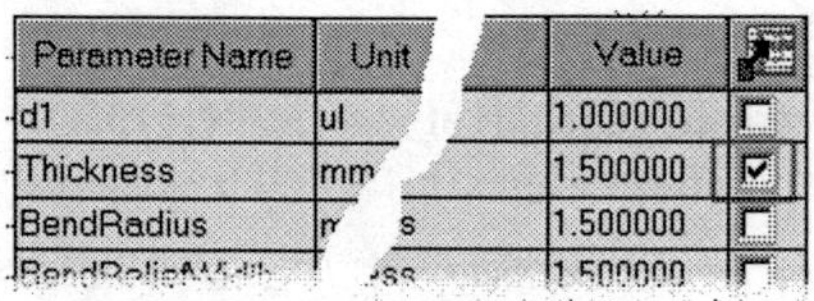

Figure 1-129

4 Click Done in the Parameters dialog box.

5 Save *SM-Trans01.ipt*.

6 Return to the assembly.

Derive a Second Component

1 Edit *SM-Trans02.ipt* in-place in the assembly.

2 Click in the Sheet Metal Panel Bar, and select Features from the menu.

3 Click the Derived Component tool.

4 In the Open dialog box:

   a. Select *SM-Trans01.ipt.*

   b. Click Open.

5 In the Derived Part dialog box:

   a. Click the Body as Work Surface.

   b. Expand the Surfaces node.

The original loft is a surface in *SM-Trans01.ipt.* You do not need to separately derive the original lofted part into the new sheet metal part.

NOTE The Exported Parameters node indicates that some parameters (Thickness) will be imported.

---

6 Click OK.

7 From the Browser, turn off the visibility of the SM-Trans01 part.

Your assembly should match the one shown in Figure 1-130.

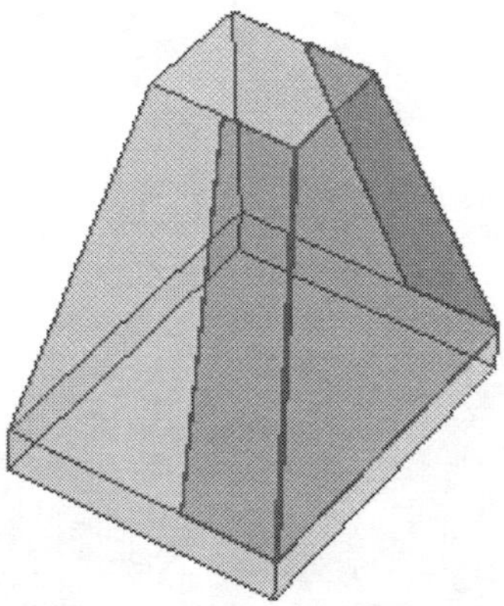

Figure 1-130

8 Expand the *SM-Trans01.ipt* derived component node in *SM-Trans02.ipt.*

NOTE The derived component contains both the surface body of the original loft, and the surface body of the first sheet metal part.

9 From the Standard toolbar, click the Parameters tool.

NOTE Since all sheet metal parts contain a model parameter named Thickness, the derived parameter is imported as Thickness_1.

10 Click Done in the Parameters dialog box.

The sheet metal face features of the new part will reference both surface bodies. Since you are not projecting geometry across part boundaries, no adaptive relationships are generated.

Add Sheet Metal Faces

1 From the Sheet Metal Panel Bar, click the Styles tool.
2 In the Sheet Metal Styles dialog box:
   a. Enter **Thickness_1** in the Thickness edit box.
   b. Click Save.
   c. Click Done.
3 Start a new sketch on the face opposite to the first sheet metal part, as shown in Figure 1-131.

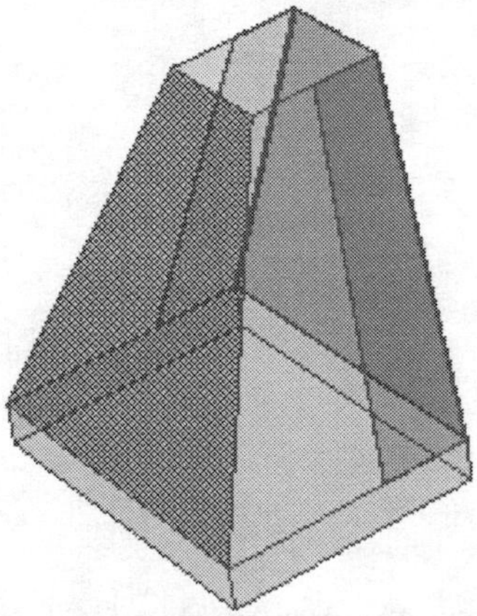

Figure 1-131

4 Exit from the sketch.

5 From the Sheet Metal Panel Bar, click the Face tool. Ensure that the sheet metal face extrudes away from the surface body.
6 Click OK.
7 Start a new sketch on the face highlighted in Figure 1-132.

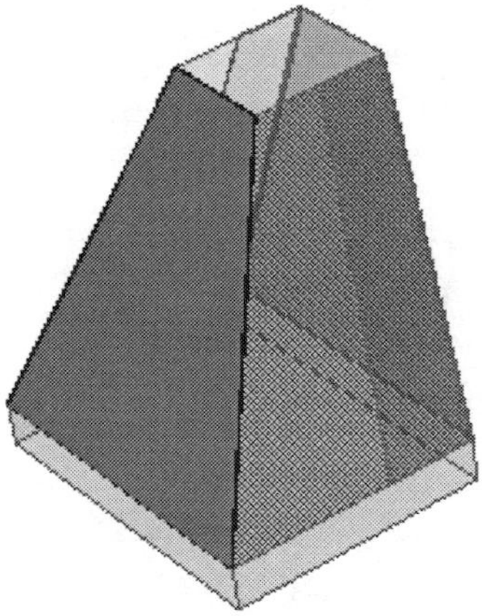

Figure 1-132

8 From the Sketch Panel Bar, click the Project Geometry tool.
9 Project the edge of the SM-Trans01 surface body as shown in Figure 1-133.
10 Sketch a new line parallel to this projected edge, coincident to the two horizontal edges as shown.
11 Add a 10 mm dimension between the two lines.

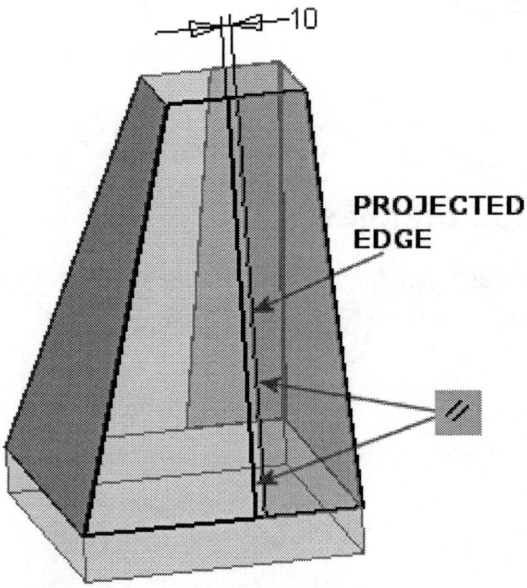

Figure 1-133

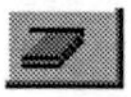

12 Exit the sketch.

13 Click the Face tool on the Sheet Metal Panel Bar.

14 Select the closed profile adjacent to the existing face.

15 In the Face dialog box:

 a. Click the Select Edge tool.

 b. Click on the adjacent edge of the existing face.

 c. Click OK in the Face dialog box

Your part should match the one shown in Figure 1-134.

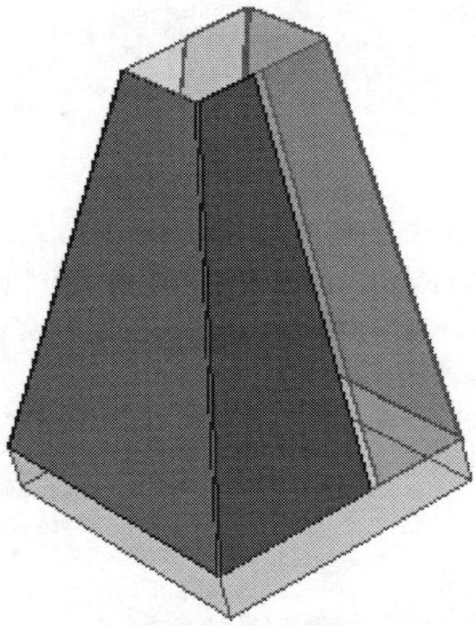

Figure 1-134

Add Sheet Metal Flanges

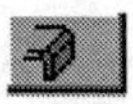

1 From the Sheet Metal Panel Bar, click the Flange tool.

2 Click the edge of the new face highlighted in Figure 1-135.

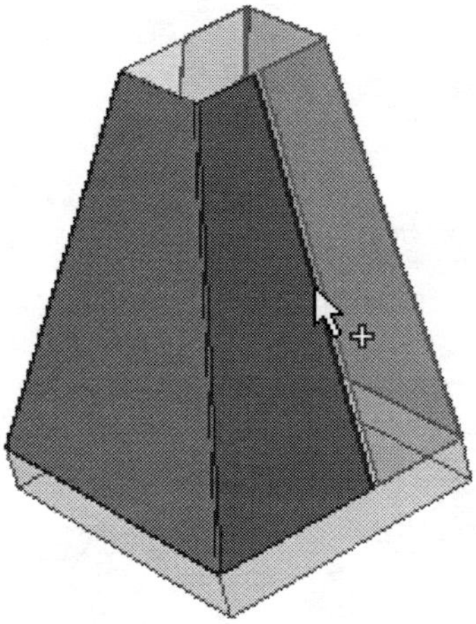

Figure 1-135

3 In the Flange dialog box:

   a. Enter **30 mm** in the Distance edit box.

   b. Enter **0 deg** in the Angle edit box.

   c. Click OK.

The preview and resulting flange features are shown in Figure 1-136. The view is from the base of the transition, looking parallel to the selected edge from Figure 1-135.

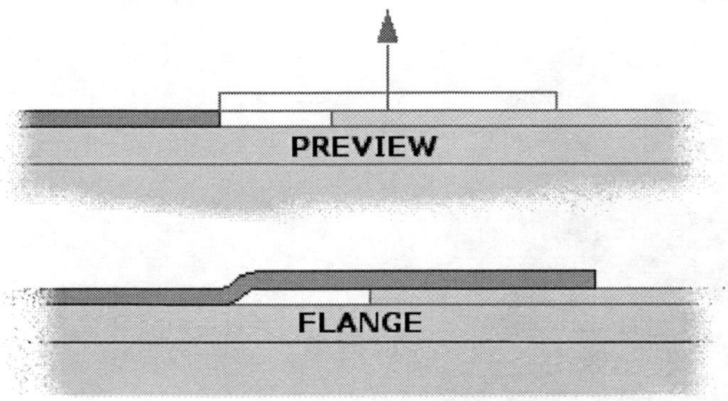

Figure 1-136

4 Create a similar face and jogged flange on the opposite side of the transition.

Your part should match the one shown in Figure 1-137.

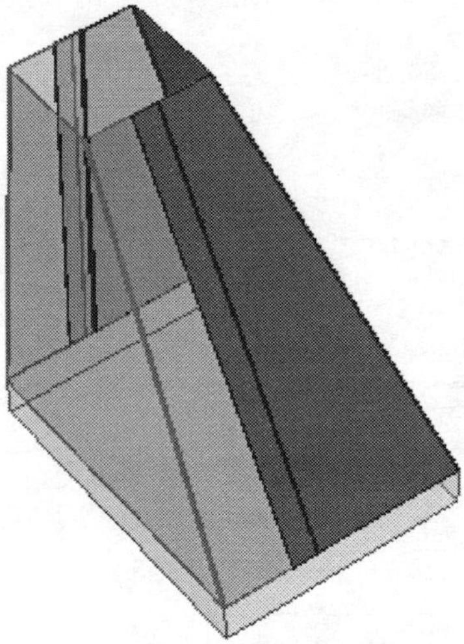

Figure 1-137

5 Expand the SM-Trans01 derived component node under *SM-Trans02.ipt*.
6 Turn off the visibility of both derived surface bodies.
7 Edit *SM-Trans01.ipt*.
8 Turn off the visibility of the derived surface body in that part.
9 Return to the top-level assembly.

The transition (with modified colors) should match the one shown in Figure 1-138.

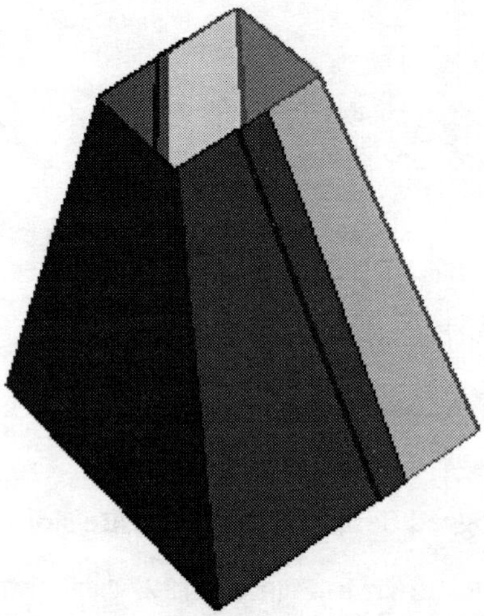

Figure 1-138

10 Finally, edit *LoftTrans.ipt* and change the dimensions of one of the loft sketches.

11 Return to the assembly. Both sheet metal parts update to match the changes.

An example is shown in Figure 1-139.

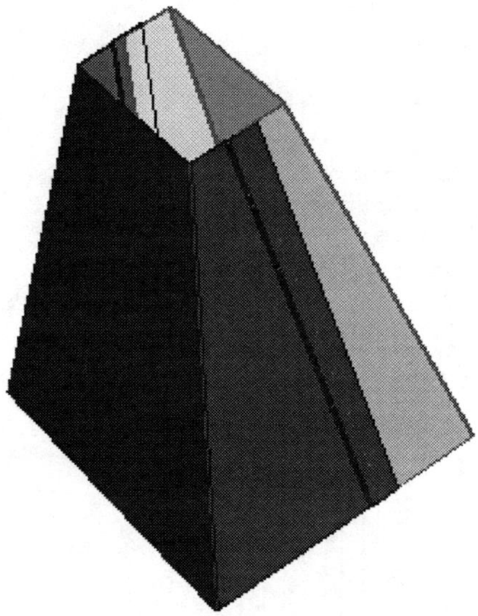

Figure 1-139

You can continue to create faces for each part from the rectangular extrusion from the LoftTrans part. It would be difficult to create a set of orthogonal flanges from the existing sheet metal faces, since every face is at a different angle from the base.

The techniques illustrated in this exercise produce robust, associative assembly models without the use of assembly constraints or adaptive relationships. The assembly components remain fully associative through the commonly derived surfaces, sketches, and parameters.

12 Close the file without saving changes.

End of exercise.

Applying Your Skills

Using the knowledge you gained through this course, create the sheet metal part shown in Figure 1-140. As part of the exercise, create a drawing that details both the flat and folded states of the part.

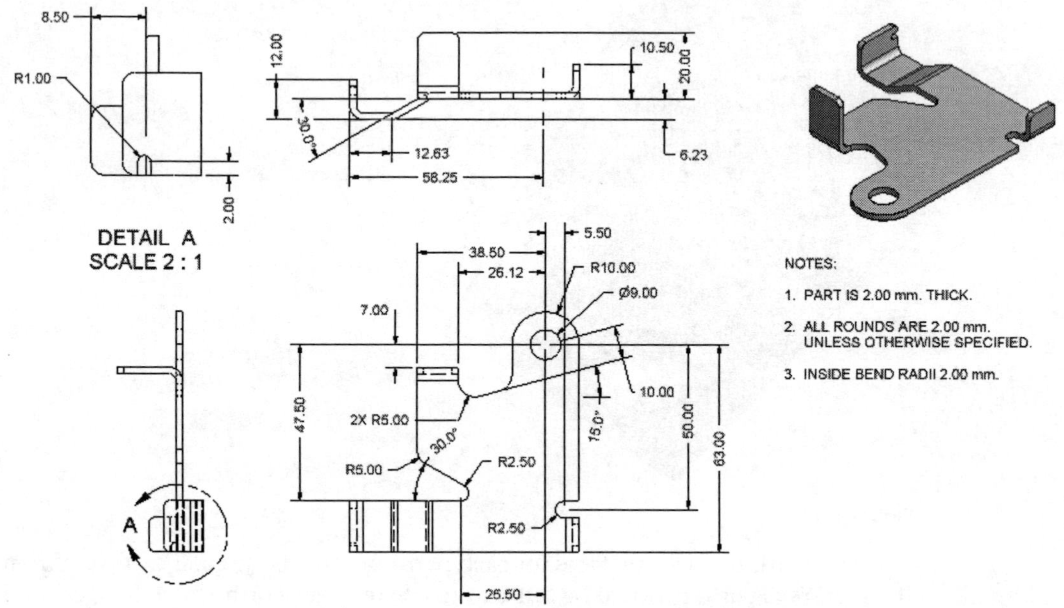

Figure 1-140

Course Summary

This course focused on specific techniques and methods that can be used to create sheet metal components using Autodesk Inventor.

You should now be able to successfully design sheet metal models using various construction techniques that were introduced to you throughout the course.

Some of these included

- In–place assembly design
- Face–flange construction
- Shelled solids with ripped corners
- Turning a flat pattern into a folded part
- Working with imported geometry

In addition, you were also introduced to techniques that help automate the sheet metal tools using sheet metal iFeatures and the PunchTool within the sheet metal environment.